A Gloss
Plastics Terminology
in 5 Languages

5th edition, with CD-ROM

English Deutsch Français
Español Italiano

Edited by W. Glenz

in collaboration with
C. Celata, H. Guyot, I. Linazisoro

HANSER

Hanser Publishers, Munich
Hanser Gardner Publications, Inc., Cincinnati

Editor
Dr. Wolfgang Glenz, *Kunststoffe / plast europe*
Der Breite Michelsweg 3, D-67547 Worms, Germany

Distributed in the USA and in Canada by
Hanser Gardner Publications, Inc.
Cincinnati, Ohio 45244-4090, USA
Fax: +1 (513) 527-8950
Internet: http://www.hansergardner.com

Distributed in all other countries by
Carl Hanser Verlag
Postfach 860420, 81631 München, Germany
Fax: +49 (89) 98 12 64
Internet: http://www.hanser.de

Die Deutsche Bibliothek – CIP-Einheitsaufnahme

Ein Titeldatensatz für diese Publikation ist bei Der Deutschen Bibliothek
erhältlich.

Library of Congress Cataloging-in-Publication Data

A CIP record for this book is available from the Library of Congress.

ISBN 1-56990-328-X (Hanser Gardner)
ISBN 1-446-21606-5 (Hanser)

© Carl Hanser Verlag, Munich 2001
Production Management: Oswald Immel
Typeset, printed and bound in Germany by Sommer-Druck
Coverdesign: MCP · Susanne Kraus GbR, Holzkirchen, Germany

Preface to the Fifth Edition

This revised and enlarged edition of the glossary contains 1795 words connected with chemistry, properties, testing, and the technology of plastics; i.e. 15 % more entries than the previous edition. The editorial team incorporated only a limited number of new terms to make sure that the glossary remains a quick-reference volume.

The concept of this glossary remained the same as in the first edition: a word-to-word translation without further explanation or interpretation. With this in mind, the user must be aware that this glossary is no substitute for a concise technical dictionary.

The glossary is particularly valuable for a quick relation of technical terms from a foreign language to the respective mother tongue. Even for translations of short technical texts, brochures, etc., this collection of technical terms usually is sufficient.

From the beginning, Dr. H. Guyot (editor-in-chief, Usipass, Paris/ France) has provided the French translations. The Italian terms were originally compiled by Dr. R. Marchelli; since the fourth edition, C. Celata (managing director, Assocomaplast, Assago/ Italy) has joined the editorial team. The Spanish translations were compiled by M. Santolaria Mur for the first editions; this time, I. Linazisoro (editor-in-chief, Plásticos Universales and Plast-univers, Bilbao/Spain) is responsible for the Spanish terms. The English terms were proofread by Raymond Brown, Newbridge/ Ireland. Jutta Graf cross-checked the translations and arranged the glossary in its present form.

I would like to express my thanks to all my colleagues in this editorial team for their continuous support and advice.

This Glossary is also available in electronic format – a CD-ROM is included on the back cover. Detailed information on page 385.

Users are welcome to submit comments and suggestions for improvements.

Wolfgang Glenz

Preface to the First Edition

With this little glossary of plastics terminology we hope to breach the language barriers within the international plastics community.

In trying to serve plastics engineers in different countries, the idea was born to compile a glossary of plastics terminology in five languages. Although we are aware of the problems involved in such a word-to-word translation of terms without further explanation or interpretation, it was nevertheless felt that such a "vademecum" would be useful for the plastics technologist.

October 1990   The Editors

Contents

Inhalt   Table des Matières   Indice   Contenuto

Nr.	English	Deutsch
1	**ablative**	ablativ
2	**abrasion**	Abrieb *m*
3	**abrasive**	abrasiv
4	**absorption**	Absorption *f*
5	**accelerator**	Beschleuniger *m*
6	**accumulator**	Akkumulator *m*
7	**accumulator head**	Kopfspeicher *m;* Speicherkopf *m*
8	**acidity**	Säuregrad *m*
9	**acoustic properties**	akustische Eigenschaften *fpl*
10	**acrylic glass** *(PMMA)*	Acrylglas *n*
11	**acrylic resin**	Acrylharz *n*
12	**acrylic rubber**	Acrylkautschuk *m*
13	**acrylonitrile butadiene styrene copolymers** *(ABS)*	Acrylnitril-Butadien-Styrol-Copolymere *npl*
14	**activation**	Aktivierung *f*
15	**adapter**	Passstück *n*
16	**additive**	Additiv *n*

Français	*Español*	*Italiano*
ablatif	ablativo	ablativo
abrasion *f*	abrasión *f*	abrasione *f*
abrasif	abrasivo	abrasivo
absorption *f*	absorción *f*	assorbimento *m*
accélérateur *m*	acelerador *m*	acceleratore *m*
accumulateur *m*	acumulador *m*	accumulatore *m*
tête *f* d' accumulation	cabezal *m* acumulador	testa *f* di accumulo
taux *m* d'acidité	grado *m* de acidez	acidità *f*
propriétés *fpl* acoustiques	propiedades *mpl* acústicas	proprietà *fpl* acustiche
verre *m* acrylique	vidrio *m* acrílico	vetro *m* acrilico
résine *f* acrylique	resina *f* acrílica	resina *f* acrilica
caoutchouc *m* acrylique	caucho *m* acrílico	gomma *f* acrilica
copolymères *mpl* d'acrylonitrile-butadiène-styrène	copolímeros *mpl* acrilonitrilo-butadieno-estireno	copolimeri *mpl* acrilonitrile butadiene stirene
activation *f*	activación *f*	attivazione *f*
adaptateur *m*	adaptador *m*	adattatore *m*
additif *m*	aditivo *m*	additivo *m*

Nr.	English	Deutsch
17	**adhere, to**	haften
18	**adhesion**	Adhäsion *f*
19	**adhesive**	Klebstoff *m*
20	**adhesive film**	Klebefolie *f*
21	**adhesive strength**	Haftfestigkeit *f*
22	**adhesive tape**	Klebeband *n*
23	**adjusting screw**	Justierschraube *f*
24	**adsorption**	Adsorption *f*
25	**after-shrinkage**	Nachschwindung *f*
26	**ageing**	Alterung *f*
27	**ageing resistance**	Alterungsbeständigkeit *f*
28	**agglomerate**	Agglomerat *n*
29	**air bubble**	Luftblase *f*
30	**air cushioning film**	Luftpolsterfolie *f*
31	**air ejector**	Luftauswerfer *m*
32	**air inlet**	Lufteintritt *m*
33	**air knife**	Luftrakel *m*

Français	Español	Italiano
adhérer	adherir	aderire
adhésion *f*	adhesión *f*	adesione *f*
adhésif *m;* colle *f*	adhesivo *m;* cola *f*	adesivo *m*
feuille *f* adhésive	película *f* adhesiva	pellicola *f* adesiva
force *f* d'adhésion	adhesión *f*	forza *f* di adesione
ruban *m* adhésif	cinta *f* adhesiva	nastro *m* adesivo
vis *f* de réglage	tornillo *m* de ajuste	dado *m* di regolazione
adsorption *f*	adsorción *f*	assorbimento *m*
post-retrait *m*	contracción *f* posterior de moldeo	post-ritiro *m*
vieillissement *m*	envejecimiento *m*	invecchiamento *m*
tenue *f* au vieillissement	resistencia *f* al envejecimiento	resistenza *f* all'invecchiamento
agglomérat *m*	aglomerado *m*	agglomerato *m*
bulle *f* d'air	burbuja *f* de aire	bolla *f* d'aria
film *m* de coussin d'air	película *f* de celdillas de aire	film *m* cuscino d'aria
éjecteur *m* à air	expulsor *m* de aire	estrattore *m* ad aria
entrée *f* d'air	entrada *f* del aire	griglia *f* presa d'aria
racle *f* en l'air	cuchilla *f* flotante	lama *f* d'aria

Nr.	English	Deutsch
34	**air nozzle**	Luftdüse *f*
35	**air outlet**	Luftaustritt *m*
36	**aldehyde resin**	Aldehydharz *n*
37	**aliphatic**	aliphatisch
38	**alkyd resin**	Alkydharz *n*
39	**alloy**	Legierung *f*
40	**allyl resin**	Allylharz *n*
41	**alternating stress**	Wechselbeanspruchung *f*
42	**aluminium trihydrate**	Aluminiumtrihydrat *n*
43	**aminoplasts**	Aminoplaste *mpl*
44	**amorphous**	amorph
45	**ancillary equipment**	Zusatzeinrichtung *f*
46	**angular adjustment**	Schrägverstellung *f*
47	**anionic polymerisation**	anionische Polymerisation *f*
48	**anisotropy**	Anisotropie *f*
49	**annealing**	Tempern *n*

Français	*Español*	*Italiano*
buse *f* d'air	boquilla *f* de aire	ugello *m* ad aria
sortie *f* d'air	salida *f* de aire	griglia *f* uscita aria
résine *f* aldéhyde	resina *f* aldehídica	resina *f* aldeidica
aliphatique	alifático	alifatico
résine *f* alkyde	resina *f* alquídica	resina *f* alchidica
alliage *m*	aleación *f*	lega *f*
résine *f* allylique	resina *f* alílica	resina *f* allilica
contrainte *f* alternée	esfuerzo *m* alternante	sollecitazione *f* alternativa
trihydrate *m* d' alumine	trihidrato *m* de alúmina	triidrato *m* di alluminio
aminoplastes *mpl*	aminoplásticos *mpl*	amminoplasti *mpl*
amorphe	amorfo	amorfo
équipement *m* périphérique	equipos *mpl* auxiliares	attrezzatura *f* ausiliaria
réglage *m* angulaire	ajuste *m* angular	regolazione *f* angolare
polymérisation *f* anionique	polimerización *f* aniónica	polimerizzazione *f* anionica
anisotropie *f*	anisotropía *f*	anisotropia *f*
recuisson *m*	recocido *m*	ricottura *f*

Nr.	English	Deutsch
50	**annular piston**	Ringkolben *m*
51	**antiblocking agent**	Antiblockmittel *n*
52	**anticorrosive**	Korrosionsschutz *m*
53	**antifogging agent**	Antibeschlagmittel *n*
54	**antioxidant**	Antioxidans *n*
55	**antiozonant**	Ozonschutzmittel *n*
56	**antistatic agent**	Antistatikum *n*
57	**apparent density**	Rohdichte *f;* Schüttdichte *f*
58	**applicator roll**	Auftragwalze *f*
59	**aramid fibre**	Aramidfaser *f*
60	**arc resistance**	Lichtbogenfestigkeit *f*
61	**ash content**	Aschegehalt *m*
62	**assembly**	Montage *f*
63	**atactic**	ataktisch
64	**austenitic**	austenitisch

Français	*Español*	*Italiano*
piston *m* annulaire	pistón *m* anular	pistone *m* anulare
agent *m* antibloquant	agente *m* antibloqueante	agente *m* antibloccante
protection *f* anticorrosive	protección *f* anticorrosiva	anticorrosivo *m*
agent *m* antifogging	agente *m* antivaho	agente *m* antiappannante
antioxydant *m*	antioxidante *m*	antiossidante *m*
anti-ozone *m*	antiozonante *m*	antiozonante *m*
agent *m* antistatique	agente *m* antiestático	agente *m* antistatico
densité *f* apparente	densidad *f* aparente	densità *f* apparente
rouleau *m* enducteur	rodillo *m* aplicador	rullo *m* per spalmatura
fibre *f* aramide	fibra *f* de aramida	fibra *f* aramidica
résistance *f* à l'arc	resistencia *f* al arco	resistenza *f* all' arco
taux *m* de cendres	contenido *m* de cenizas	contenuto *m* di ceneri
assemblage *m*	montaje *m*	assemblaggio *m*; gruppo *m*
atactique	atáctico	atattico
austénitique	austenítico	austenitico

Nr.	English	Deutsch
65	**automatic press**	Pressautomat *m*
66	**autoxidation**	Autoxidation *f*
67	**axial stress**	Axialspannung *f*
68	**back injection**	Hinterspritzen *n*
69	**back pressure**	Staudruck *m*
70	**backing plate**	Aufspannkörper *m*
71	**bagging unit**	Absackanlage *f*
72	**ball bearing**	Kugellager *n*
73	**ball check valve**	Kugelrückschlagventil *n*
74	**ball indentation hardness**	Kugeldruckhärte *f*
75	**ball mill**	Kugelmühle *f*
76	**Banbury mixer** *(internal mixer with plug)*	Banbury-Mischer *m*
77	**band heater**	Heizungsband *n*
78	**bar code**	Strichcode *m*
79	**barrel**	Zylinder *m*

Français	*Español*	*Italiano*
presse *f* automatique	prensa *f* automática	pressa *f* automatica
autooxydation *f*	autooxidación *f*	autossidazione *f*
contrainte *f* axiale	esfuerzo *m* axial	sollecitazione *f* assiale
rétroinjection *f*	inyección *f* trasera	retroiniezione *f*
contre-pression *f*	contrapresión *f*	pressione *f* di rincalzo o di mantenimento
plateau *m* de fixation	placa *f* de sujección	piastra *f* di supporto
ensacheuse *f*	planta *f* de ensacado	insaccatrice *f*
roulement *m* à billes	cojinete *m* de bolas	cuscinetto *m* per spinte oblique
clapet *m* anti-retour	válvula *f* de esfera antirretorno	valvola *f* a sfera antiritorno
test *m* de dureté à la bille	dureza *f* por penetración de la bola	durezza *f* alla penetrazione della biglia
broyeur *m* de balles	molino *m* de bolas	mulino *m* a palle
mélangeur *m* Banbury	mezclador *m* Banbury	mescolatore *m* Banbury
collier *m* chauffant	banda *f* calefactora	banda *f* riscaldante
code *m* barre	código *m* de barras	codice *m* a barre
cylindre *m*	cilindro *m*	cilindro *m*; tamburo *m*

Nr.	English	Deutsch
80	**barrel jacket**	Zylindermantel *m*
81	**barrier film**	Sperrschichtfolie *f*
82	**barrier layer**	Sperrschicht *f*
83	**barrier plastic**	Barrierekunststoff *m*
84	**barrier properties**	Barriereeigenschaften *fpl*
85	**barrier resins**	Sperrschichtpolymere *npl*
86	**barrier screw**	Barriereschnecke *f*
87	**base frame**	Grundgestell *n*
88	**bead**	Schweißraupe *f;* Schweißwulst *m*
89	**bead polymerisation**	Perlpolymerisation *f*
90	**bending**	Biegen *n*
91	**bevel gear tooth system**	Kegelradverzahnung *f*
92	**bevel gear transmission**	Kegelradgetriebe *n*
93	**biaxial**	biaxial

Français	*Español*	*Italiano*
enveloppe *f* de cylindre	camisa *f* del cilindro	camicia *f* del cilindro
film *m* barrière	película *f* barrera	pellicola *f* barriera
couche *f* barrière	estrato *m* barrera	strato *m* barriera
matière *f* plastique barrière	plástico *m* barrera	materia *f* plastica barriera
propriétés *fpl* barrières	propiedades *fpl* de barrera	proprietà *fpl* barriera
résines *fpl* barrières	capas *fpl* barrera de polímero	resine *fpl* barriera
vis *f* barrière	husillo *m* barrera	vite *f* barriera
châssis *m*	bancada *f*	telaio *m* base
cordon *m* de soudure	cordón *m* de soldadura	cordone *m* di saldatura
polymérisation *f* en suspension	polimerización *f* en perla	polimerizzazione *f* in perle
cintrage *m*	curvado *m*	curvatura *f*
dents *f* à engrenage en biseaux	sistema *m* dentado de engranaje cónico	sistema *f* a ingranaggio conico dentato
transmission *m* par engrenage à biseaux	transmisión *f* de piñón cónico	trasmissione *f* a ingranaggio conico
biaxial	biaxial	biassiale

Nr.	English	Deutsch
94	**biaxially oriented**	biaxial orientiert
95	**bimetallic**	bimetallisch
96	**binder**	Bindemittel *n*
97	**biocide**	Biozid *m*
98	**biocompatibility**	Bioverträglichkeit *f*
99	**biodegradable plastics**	biologisch abbaubare Kunststoffe *mpl*
100	**biodegradation**	biologische Abbaubarkeit *f*
101	**biopolymer**	Biopolymer *n*
102	**biostabiliser**	Biostabilisator *m*
103	**birefringence**	Doppelbrechung *f*
104	**blade** *(of a mixer)*	Schaufel *f*
105	**bleeding**	Ausbluten *n*
106	**blend**	Blend *n;* Mischung *f*
107	**blind hole**	Sackloch *n*
108	**blister package**	Blisterverpackung *f*
109	**block copolymer**	Block-Copolymer *n*

Français	*Español*	*Italiano*
biorienté	orientado biaxialmente; biorientado	orientato biassialmente
bimétallique	bimetálico	bimetallico
liant *m*	aglutinante *m*	legante *m*
biocide *m*	biocida *m*	biocida *m*
biocompatibilité *f*	biocompatibilidad *f*	biocompatibilità *f*
plastiques *mpl* biodégradables	plásticos *mpl* biodegradables	materie *fpl* plastiche biodegradabili
biodégradation *f*	biodegradación *f*	biodegradazione *f*
biopolymère *m*	biopolímero *m*	biopolimero *m*
biostabilisant *m*	bioestabilizador *m*	biostabilizzante *m*
biréfringence *f*	birrefringencia *f*	birifrangenza *f*
pale *f*	paleta *f*	pala *f*
épanchement *m*	sangrado *m*; goteo *m*	essudazione *f*
blend *m*	blend *m;* mezcla *f*	mescola *f*
trou *m* borgne	canal *m* ciego	foro *m* cieco
emballage *m* blister	envasado *m* blister	imballaggio *m* blister
copolymère *m* à blocs	copolímero *m* de bloque	copolimero *m* a blocchi

Nr.	English	Deutsch
110	**block polymerisation**	Blockpolymerisation *f*
111	**blocking**	Blocken *n*
112	**blooming**	Ausblühen *n;* Ausschwitzen *n*
113	**blow moulding**	Blasformen *n*
114	**blow moulding line**	Blasformanlage *f*
115	**blow-up ratio**	Aufblasverhältnis *m*
116	**blowing agent**	Treibmittel *n*
117	**blowing mandrel**	Blasdorn *m*
118	**blowing mould**	Blasformwerkzeug *n*
119	**blown film**	Blasfolie *f*
120	**blown film die**	Folienblaskopf *m*
121	**bolt**	Bolzen *m*
122	**bolt shut-off nozzle**	Bolzenverschlussdüse *f*
123	**bond, to**	kleben

Français	*Español*	*Italiano*
polymérisation *f* en blocs	polimerización *f* en bloque	polimerizzazione *f* in blocco
bloquant *m*	bloqueo *m*	bloccaggio *m*
efflorescence *f*	eflorescencia *f*	efflorescenza *f*
soufflage *m*	moldeo *m* por soplado	soffiaggio *m*
souffleuse *f*	instalación *f* de soplado	linea *f* di soffiaggio
taux *m* de gonflage	relación *f* de soplado	rapporto *m* di soffiaggio
agent *m* gonflant	agente *m* de expansión	agente *m* di espansione
mandrin *m* de soufflage	mandril *m* de soplado	mandrino *m* di soffiaggio
moule *m* de soufflage	molde *m* de soplado	stampo *m* di soffiaggio
film *m* soufflé	película *f* soplada	pellicola *f* soffiata
tête *f* de soufflage	cabezal *m* de película soplada	filiera *f* soffiaggio film
boulon *m*	bulón *m*	bullone *m*
buse *f* à fermeture boulon	boquilla *f* de cierre mediante bulón	ugello *m* a serranda
adhérer à	pegar; unir con adhesivos	incollare

Nr.	English	Deutsch
124	**bonded joint**	Klebeverbindung *f*
125	**bonding agent**	Bindemittel *n*
126	**bonding cement**	Kitt *m*
127	**bore**	Bohrung *f*
128	**boride coated**	boriert
129	**branching**	Verzweigung *f*
130	**breaker plate**	Lochscheibe *f*
131	**breaking length**	Reißlänge *f*
132	**Brinell hardness**	Brinellhärte *f*
133	**brittle**	spröde
134	**brittle fracture**	Sprödbruch *m*
135	**brittleness**	Sprödigkeit *f*
136	**bubble**	Blase *f*
137	**bubble inflation air**	Stützluft *f*
138	**buckling resistance**	Knickfestigkeit *f*
139	**bulk density**	Schüttdichte *f*

Français	*Español*	*Italiano*
joint *m* de colle	unión *f* por encolado	giunto *m* incollato
agent *m* de couplage	agente *m* ligante	agente *m* legante
ciment *m* de liaison	masilla *f*	cemento *m* legante
alésage *m*	perforación *f*	alesaggio *m*
revêtu de borure	revestido con boruro	borurato
ramification *f*	ramificación *f*	ramificazione *f*
plaque *f* du filtre	placa *f* rompedora	piastra *f* del filtro
allongement *m* à la rupture	longitud *f* de rotura	lunghezza *f* di rottura
dureté *f* Brinell	dureza *f* Brinell	durezza *f* Brinell
fragile	frágil	fragile
rupture *f* fragile	fractura *f* frágil	rottura *f* fragile
fragilité *f*	fragilidad *f*	fragilità *f*
bulle *f;* ballon *m*	burbuja *f*	bolla *f*
air *m* pour souffler la bulle	aire *m* para soplado de la burbuja	aria *f* di gonfiaggio della bolla
résistance *f* au gauchissement	resistencia *f* al pandeo	resistenza *f* alla deformazione per compressione
densité *f* du vrac	densidad *f* de la masa	densità *f* di massa

Nr.	English	Deutsch
140	**bulk moulding compound** *(BMC)*	Feuchtpressmasse *f*; BMC
141	**bulk polymerisation**	Massepolymerisation *f*
142	**burnt streaks**	Verbrennungsschlieren *fpl*
143	**bursting pressure**	Berstdruck *m*
144	**bursting strength**	Berstdruckfestigkeit *f*
145	**bush bearing**	Lagerbuchse *f*
146	**bushing**	Buchse *f*
147	**butadiene rubber**	Butadienkautschuk *m*
148	**butt joint**	Stumpfstoß *m*
149	**butt welding**	Stumpfschweißen *n*
150	**butyl rubber**	Butylkautschuk *m*
151	**cable sheathing**	Kabelummantelung *f*
152	**cadmium pigment**	Cadmiumpigment *n*

Français	Español	Italiano
BMC *m*	compuesto *m* de moldeo por compresión; BMC	massa *f* per stampaggio
polymérisation *f* en masse	polimerización *f* en masa	polimerizzazione *f* in massa
traces *fpl* de brûlures	vetas *fpl* de quemadura	segni *mpl* di bruciatura
pression *f* d'éclatement	presión *f* de reventamiento	pressione *f* di scoppio
résistance *f* à l'éclatement	resistencia *f* al reventamiento	resistenza *f* allo scoppio
douille *f* de fixation	casquillo *m* del cojinete	supporto *m* della boccola
bague *f;* douille *f*	manguito *m*	boccola *f*
caoutchouc *m* butadiène	caucho *m* de butadieno	elastomero *m* butadiene
joint *f* bout à bout	unión *f* soldada a tope	giunto *m* di testa
soudage *m* bout à bout	soldadura *f* a tope	saldatura *f* di testa
caoutchouc *m* butyle	caucho *m* de butilo	gomma *f* butile
gainage *m* de câbles	cubierta *f* protectora de cable	protezione *f* per cavi
pigment *m* de cadmium	pigmento *m* de cadmio	pigmento *m* di cadmio

Nr.	English	Deutsch
153	**caking**	Klumpenbildung *f*
154	**calender**	Kalander *m*
155	**calendering**	Kalandrieren *n*
156	**calibrate, to**	kalibrieren
157	**calibrating device** *(pipes, profiles)*	Kalibriervorrichtung *f*
158	**calibration**	Kalibrierung *f*
159	**calibration basket**	Kalibrierkorb *m*
160	**calorimetry**	Kalorimetrie *f*
161	**capillary rheometer**	Kapillarrheometer *n*
162	**capillary viscometer**	Kapillarviskosimeter *n*
163	**carbon black**	Ruß *m*
164	**carbon fibre**	Kohlenstofffaser *f*
165	**cartridge heater**	Heizpatrone *f*
166	**cartridge valve**	Sitzventil *n*
167	**cascade extruder**	Kaskadenextruder *m;* Tandemextruder *m*

Français	*Español*	*Italiano*
formation *f* de grumeaux	formación *f* de grumos	formazione *f* dei grumi
calandre *f*	calandra *f;* calandria *f*	calandra *f*
calandrage *m*	calandrado *m*	calandratura *f*
calibrer	calibrar	calibrare
système *m* de calibrage	calibrador *m*	calibratore *m*
calibrage *m*	calibrado *m*	calibratura *f*
bloc *m* de calibrage	jaula *f* de calibrador	blocco *m* di calibrazione
calorimétrie *f*	calorimetría *f*	calorimetria *f*
rhéomètre *m* capillaire	reómetro *m* capilar	reometro *m* capillare
viscosimètre *m* capillaire	viscosímetro *m* capilar	viscosimetro *m* capillare
noir *m* de carbone	negro *m* de carbono	nerofumo *m*
fibre *f* de carbone	fibra *f* de carbono	fibra *f* di carbonio
cartouche *f* chauffante	cartucho *m* calefactor	riscaldatore *m* a cartuccia
valve *f* de cartouche	válvula *f* de cartucho	valvola *f* a cartuccia
extrudeuse *f* en cascade	extrusora *f* en cascada	estrusore *m* a cascata

Nr.	English	Deutsch
168	**case-hardened steel**	Einsatzstahl *m*
169	**case-hardening**	Einsatzhärtung *f*
170	**cast film**	Gießfolie *f*
171	**cast film extrusion**	Foliengießen *n*
172	**cast film line**	Foliengießanlage *f*
173	**cast resin**	Gießharz *n*
174	**casting**	Gießverfahren *n*; Gießen *n*
175	**catalysed lacquers**	Reaktionslacke *mpl*
176	**catalyst**	Katalysator *m*
177	**caterpillar take-off**	Raupenabzug *m*
178	**cationic polymerisation**	kationische Polymerisation *f*
179	**cavity**	Kavität *f*; Formnest *n*
180	**cavity plate**	Gesenk *n*; Matrize *f*
181	**cavity pressure**	Werkzeuginnendruck *m*

Français	*Español*	*Italiano*
acier *m* cémenté	acero *m* de cementación	acciaio *m* da cementazione
cémentation *f*	cementación *f*	cementazione *f*
film *m* coulé	película *f* colada	pellicola *f* colata
extrusion *f* de film coulé (ou cast)	extrusión *f* de película colada	estrusione *f* di film per colata
ligne *f* de film à plat	instalación *f* de película colada	linea *f* per film colato
résine *f* de coulée	resina *f* de colada	resina *f* per colata
coulée *f*	proceso *m* por colada	colata *m*
peintures *fpl* réactives	lacas *fpl* de reacción	vernici *fpl* catalizzate
catalyseur *m*	catalizador *m*	catalizzatore *m*
tireuse *f* à chenilles	arrastre *m* por orugas	dispositivo *m* di tirata a cingoli
polymérisation *f* cationique	polimerización *f* catiónica	polimerizzazione *f* cationica
empreinte *f;* cavité *f*	cavidad *f*	cavità *f;* impronta *f*
matrice *f*	placa *f* de la cavidad; matriz *f*	stampo *m* femmina
pression *f* interne du moule	presión *f* interior del molde	pressione *f* nello stampo

Nr.	English	Deutsch
182	**cavity temperature**	Werkzeugtemperatur *f*
183	**cavity wall**	Formnestwand *f*; Werkzeugwand *f*
184	**cellophane**	Zellglas *n*
185	**cellular plastic**	Schaumkunststoff *m*
186	**cellulose**	Cellulose *f*
187	**cellulose acetate**	Celluloseacetat *n*
188	**cellulose acetobutyrate**	Celluloseacetobutyrat *n*
189	**cellulose nitrate**	Cellulosenitrat *n*
190	**cellulose propionate**	Cellulosepropionat *n*
191	**centering flange**	Zentrierflansch *m*
192	**centering screw**	Zentrierschraube *f*
193	**centrifugal casting**	Schleuderverfahren *n*
194	**centrifugal deposition**	Ausschleuderung *f*

Français	*Español*	*Italiano*
température *f* de l'empreinte	temperatura *f* de la cavidad del molde	temperatura *f* nella cavità
paroi *f* d'empreinte	pared *f* de la cavidad	parete *f* dell'impronta
cellophane *f*	celofán *f*	cellofane *m*
plastique *m* cellulaire, mousse *f*	plástico celular *m*	plastica *f* espansa
cellulose *f*	celulosa *f*	cellulosa *f*
acétate *m* de cellulose	acetato *m* de celulosa	acetato *m* di cellulosa
acétobutyrate *m* de cellulose	acetobutirato *m* de celulosa	acetobutirrato *m* di cellulosa
nitrate *m* de cellulose	nitrato *m* de celulosa	nitrato *m* di cellulosa
propionate *m* de cellulose	acetopropionato *m* de celulosa	aceto propionato *m* di cellulosa
bride de centrage *f*	brida de centrado *f*	flangia *f* di centraggio
vis de centrage *f*	husillo *m* de centrado	vite *f* di centraggio
coulée *f* par centrifugation	moldeo *m* por centrifugación	colata *f* per centrifugazione
dépôt *m* par centrifugation	deposición *f* centrífuga	deposito *m* da centrifugazione

Nr.	English	Deutsch
195	**ceramic injection moulding**	Keramikspritzguss *m*
196	**chain conformation**	Kettenkonformation *f*
197	**chain folding**	Kettenfaltung *f*
198	**chain length** *(of a macromolecule)*	Kettenlänge *f*
199	**chain propagation**	Kettenwachstum *n*
200	**chain scission**	Kettenspaltung *f*
201	**chain segment**	Kettensegment *n*
202	**chain termination**	Kettenabbruch *m*
203	**chain transfer**	Kettenübertragung *f*
204	**chalk**	Kreide *f*
205	**chalking**	Kreiden *n*
206	**channel volume** *(of a screw)*	Gangvolumen *n*
207	**channel width** *(of a screw)*	Gangbreite *f*

Français	*Español*	*Italiano*
injection *f* de céramique	moldeo *m* por inyección de cerámica	stampaggio *m* a iniezione di materiali ceramici
conformation *f* de la chaîne	conformación *f* de la cadena	conformazione *f* a catena
chaînes *fpl* enchevêtrées	cadena *f* plegada	piega *f* della catena
longueur *f* de chaîne	longitud *f* de cadena	lunghezza *f* della catena
propagation *f* de la chaîne	crecimiento *m* de la cadena	allungamento *m* della catena
coupure *f* de chaîne	escisión *f* de la cadena	rottura *f* della catena
segment *m* de chaîne	segmento *m* molecular	segmento *m* della catena
terminaison *f* de chaîne	terminación *f* de cadena	terminazione *f* della catena
transfert *m* de chaîne	transferencia *f* de cadena	traferitore *m* della catena
craie *f*	creta *f*	creta *f*
farinage *m*	caleo *m*	sfarinamento *m*
volume *m* du canal	volumen *m* del paso	volume *m* del canale
largeur *f* des canaux	anchura *f* del paso	ampiezza *f* del canale

Nr.	English	Deutsch
208	**check valve**	Rückschlagventil *n*
209	**chelator**	Chelator *m*
210	**chemical resistance**	Chemikalienbeständigkeit *f*
211	**chill roll**	Kühlwalze *f*
212	**chill roll casting**	Chillroll-Verfahren *n*
213	**chlorinated hydrocarbons**	chlorierte Kohlenwasserstoffe *mpl*
214	**choke bar**	Staubalken *m*
215	**chopped glass fibres**	Kurzglasfasern *fpl*
216	**chopped strands**	Schnittglasfasern *fpl*
217	**chute**	Rutsche *f*
218	**circulating air drying oven**	Umlufttrockner *m*
219	**clamping cylinder**	Schließzylinder *m*
220	**clamping force**	Schließkraft *f*
221	**clamping jaw**	Spannbacke *f*

Français	*Español*	*Italiano*
soupape *f* de retenue	válvula *f* de retención	valvola *f* di ritegno
agent *m* chélatant	quelante *m*	agente *m* chelante
résistance *f* chimique	resistencia *f* química	resistenza *f* chimica
cylindre *m* refroidisseur	cilindro *m* de enfriamiento	cilindro *m* di raffreddamento
procédé *m* au cylindre refroidisseur	colada *m* sobre rodillo frío	fusione *f* del rullo di raffreddamento
hydrocarbures *mpl* chlorés	hidrocarburos *mpl* clorados	idrocarburi *mpl* clorurati
barre *f* de régulation	barra *f* de regulación	barra *f* di regolazione
fibres *fpl* de verre courtes	fibras *fpl* de vidrio cortas	fibre *fpl* di vetro tagliata
fibres *fpl* coupées	fibras *fpl* cortadas	fibre *fpl* tagliate
glissière *f*	guía *f* de canalización de piezas	scivolo *m*
four *m* de séchage à circulation d'air chaud	secador *m* con circulación de aire	forno *m* di essiccamento a circolazione d'aria
cylindre *m* de fermeture	cilindro *m* de cierre	cilindro *m* di chiusura
force *f* de fermeture	presión *f* de cierre	forza *f* di chiusura
mâchoire *f* de serrage	mordaza *f* de sujeción	ganascia *f*

Nr.	English	Deutsch
222	**clamping plate**	Aufspannplatte *f*
223	**clamping unit**	Schließeinheit *f*
224	**clarifying agent**	Klarifizierungsmittel *n*
225	**classification**	Klassifizierung *f*
226	**closed-cell** *(foam)*	geschlossenzellig
227	**closed-loop control**	geschlossener Regelkreis *m*
228	**coagulation**	Koagulation *f*
229	**coathanger die**	Kleiderbügeldüse *f*
230	**coating**	Beschichten *n;* Lackieren *n*
231	**coating compounds**	Beschichtungsmassen *fpl*
232	**coating resin**	Lackharz *n*
233	**coefficient of expansion**	Ausdehnungskoeffizient *m*
234	**coefficient of sliding friction**	Gleitreibungskoeffizient *m*
235	**coextrusion**	Coextrusion *f*

Français	*Español*	*Italiano*
plateau *m* de fermeture	plato *m* de cierre	piastra *f* di chiusura
unité *f* de fermeture	unidad *f* de cierre	unità *f* di chiusura
agent *m* clarifiant	agente *m* clarificante	agente *m* chiarificante
classification *f*	clasificación *f*	classificazione *f*
cellules *fpl* fermées	células *fpl* cerradas	celle *fpl* chiuse
régulation *f* en boucle fermée	control *m* por bucle cerrado	controllo *m* a circuito chiuso
coagulation *f*	coagulación *f*	coagulazione *f*
filière *f* en cintre	boquilla *f* en forma de "percha"	bocchettone *m* a attaccapanni
revêtement *m;* enduction *f*	revestimiento *m*	spalmatura *f;* rivestimento *m*
produits *mpl* d'enduction	masas *fpl* de revestimiento	mescole *fpl* per rivestimento
résine *f* pour peintures	resina *f* para barniz	resina *f* per vernici
coefficient *f* de dilatation	coeficiente *m* de dilatación	coefficiente *m* d'espansione
coefficient *m* de friction	coeficiente *m* de fricción en deslizamiento	coefficiente *m* di frizione allo scorrimento
coextrusion *f*	coextrusión *f*	coestrusione *f*

Nr.	English	Deutsch
236	**coextrusion line**	Coextrusionsanlage *f*
237	**coil**	Spule *f*
238	**coil model**	Knäuelmodell *n*
239	**cold drawing**	Kaltverstrecken *n*
240	**cold forming**	Kaltumformen *n*
241	**cold mixing**	Kaltmischung *f*
242	**cold runner**	Kaltkanal *m*
243	**cold-runner mould**	Kaltkanalwerkzeug *n*
244	**cold setting**	Kalthärtung *f*
245	**cold slug**	kalter Pfropfen *m*
246	**collapsible core**	Faltkern *m*
247	**collodium cotton**	Collodiumwolle *f*
248	**colorant**	Farbmittel *n*
249	**colorimetry**	Farbmetrik *f*
250	**colour change**	Farbwechsel *m*

Français	*Español*	*Italiano*
ligne *f* de coextrusion	instalación *f* de coextrusión	linea *f* di coestrusione
pelote *f*	bobina *f*	bobina *f*
modèle *m* pelote	modelo *m* ovillado	modello *m* bobina
étirage *m* à froid	estirado *m* en frío	stiro *m* a freddo
formage *m* à froid	moldeo *m* en frío	formatura *f* a freddo
mélange *m* à froid	mezclado *m* en frío	mescola *f* a freddo
canal *m* froid	canal *m* frío	canale *m* freddo
moule *m* à canaux froids	molde *m* de canal frío	stampo *m* a canali freddi
réglage *m* à froid	endurecimiento *m* en frío	indurimento *m* a freddo
goutte *f* froide; bouchon *m* froid	tapón *m* frío; gota *f* fría	goccia *f* fredda
noyau *m* éclipsable	núcleo *m* colapsable	nucleo *m* a sprofondamento
laine *f* colloïdale	celulosa *f* nitrada	cotone *m* collodio
colorant *m*	colorante *m*	colorante *m*
colorimétrie *f*	colorímetro *m*	colorimetria *f*
changement de couleur *m*	cambio de color *m*	cambio *m* colore

Nr.	English	Deutsch
251	colour stability	Farbbeständigkeit *f*
252	coloured streaks	Farbschlieren *fpl*
253	colouring	Einfärben *n*
254	column	Säule *f;* Holm *m*
255	combustibility	Brennbarkeit *f*
256	commodities	Standard-Kunststoffe *mpl*
257	comonomer	Comonomer *n*
258	compatibiliser	Verträglichkeitsmacher *m*
259	compatibility	Verträglichkeit *f*
260	compliance	Nachgiebigkeit *f*
261	composite material	Verbundwerkstoff *m*
262	compound	Compound *n*
263	compounder	Compoundieranlage *f*
264	compounding	Compoundieren *n;* Aufbereiten *n*
265	compressibility	Kompressibilität *f*

Français	*Español*	*Italiano*
stabilité *f* de la couleur	estabilidad *f* de color	stabilità *f* del colore
traces *fpl* de couleur	vetas *fpl* coloreadas	segni *mpl* colorati
coloration *f*	coloración *f*	coloritura *f*
colonne *f*	columna *f*	colonna *f*
inflammabilité *f*	combustibilidad *f*	combustibilitá *f*
plastiques *mpl* de grande consommation	plásticos *mpl* de gran consumo	resine *fpl* di largo consumo
comonomère *m*	comonómero *m*	comonomero *m*
compatibilisant *m*	compatibilizador *m*	compatibilizzante *m*
compatibilité *f*	compatibilidad *f*	compatibilità *f*
compliance *f*	flexibilidad *f*	conformità *f*
matériau *m* composite	material *m* laminado; composite *m*	materiale *m* composito
compound *m*	compound *m;* mezcla *f*	mescola *f*
ligne *f* de compoundage	instalación *f* para preparar compuestos	mescolatore *m*
compoundage *m*	preparación *f* de compuestos	miscelazione *f*
compressibilité *f*	capacidad *f* de compresión	compressibilità *f*

Nr.	English	Deutsch
266	compression	Kompression *f*
267	compression mould	Presswerkzeug *n*
268	compression mould insert	Presswerkzeugeinsatz *m*
269	compression moulding	Pressformen *n;* Pressen *n*
270	compression section *(screw)*	Verdichtungszone *f;* Kompressionszone *f*
271	compression set	Druckverformungsrest *m*
272	compression test	Druckversuch *m*
273	compression time	Presszeit *f*
274	compressive strength	Druckfestigkeit *f*
275	compressive stress	Druckspannung *f*
276	computer aided design *(CAD)*	rechnergestütztes Konstruieren *n*
277	computer aided manufacturing *(CAM)*	rechnergestützte Fertigung *f*
278	computer integrated manufacturing *(CIM)*	rechnerintegrierte Fertigung *f*

Français	*Español*	*Italiano*
compression *f*	compresión *f*	compressione *f*
moule *m* de compression	molde *m* de compresión	stampo *m* a compressione
insert *m* de moule de compression	inserto *m* del molde de compresión	inserto *m* di stampo a compressione
moulage *m* par compression	moldeo *m* por compresión	stampaggio *m* a compressione
zone *f* de compression	zona *f* de compresión	zona *f* di compressione
compression *f* rémanente	compresión *f* residual	compressione *f* residua
test *m* de compression	ensayo *m* de compresión	prova *f* di compressione
temps *m* de compression	tiempo *m* de prensado	tempo *m* di compressione
résistance *f* à la compression	resistencia *f* a la compresión	resistenza *f* alla compressione
contrainte *f* de compression	esfuerzo *f* de compresión	sollecitazione *f* alla compressione
conception *f* assistée par ordinateur (CAO)	diseño *m* asistido por ordenador	progettazione *f* assistita dal calcolatore
fabrication *f* assistée par ordinateur (FAO)	fabricación *f* asistida por ordenador	produzione *f* assistita dal calcolatore
fabrication *f* intégrée par ordinateur (CIM)	fabricación *f* integrada por ordenador	produzione *f* integrata con il calcolatore

Nr.	English	Deutsch
279	**condensation poly-merisation**	Kondensations-polymerisation *f*
280	**conditioning**	Konditionieren *n*
281	**conditioning chamber**	Klimakammer *f*
282	**conductive plastics**	leitfähige Kunststoffe *mpl*
283	**cone gate**	Schirmanguss *m*
284	**configuration**	Konfiguration *f*
285	**conformation**	Konformation *f*
286	**conical**	konisch
287	**constitution** *(chemical)*	Konstitution *f*
288	**contact adhesion**	Kontaktkleben *n*
289	**contact compression moulding**	Kontaktpressen *n*
290	**contact pressure**	Anpressdruck *m*
291	**contamination**	Verschmutzung *f*
292	**continuous filament**	Endlosfaser *f*
293	**control**	Steuerung *f*; Regelung *f*

Français	Español	Italiano
polymérisation *f* par condensation	polimerización *f* por condensación	polimerizzazione *f* per condensazione
conditionnement *m*	acondicionamiento *m*	condizionamento *m*
enceinte *f* climatique	cámara *f* de acondicionamiento	camera *f* di condizionamento
plastiques *mpl* conducteurs	plásticos *mpl* conductivos	materie plastiche *fpl* conduttive
injection *f* en voûte	entrada *f* cónica	entrata *f* conica
configuration *f*	configuración *f*	configurazione *f*
conformation *f*	conformación *f*	conformazione *f*
conique	cónico	conico
constitution *f*	constitución *f*	costituzione *f*
collage *m* au contact	adhesión *f* por contacto	adesione *f* a contatto
moulage *m* au contact	moldeo *m* de presión por contacto	stampaggio *m* per compressione a contatto
pression *m* de contact	presión *f* de apriete	pressione *f* di contatto (nello stampaggio)
contamination *f*	contaminación *f*	contaminazione *f*
filament *m* continu	filamento *m* continuo	filamento *m* continuo
régulation *f*	control *m*	controllo *m*

Nr.	English	Deutsch
294	**control loop**	Regelkreis *m*
295	**conveying screw**	Förderschnecke *f*
296	**conveyor**	Förderanlage *f*
297	**conveyor belt**	Förderband *n*
298	**cool, to**	abkühlen; kühlen
299	**coolant**	Kühlmedium *n*
300	**cooling capacity**	Kühlleistung *f*
301	**cooling channel**	Kühlkanal *m*
302	**cooled die**	Kühldüse *f*
303	**cooling jacket**	Kühlmantel *m*
304	**cooling mixer**	Kühlmischer *m*
305	**cooling rate**	Abkühlgeschwindigkeit *f*
306	**cooling ring**	Kühlring *m*
307	**cooling time**	Kühlzeit *f*

Français	*Español*	*Italiano*
boucle *f* de régulation	control *m* por bucle	anello *m* di controllo
vis *f* transporteuse	tornillo *m* transportador	vite *f* trasportatrice
bande *f* transporteuse	transportador *m* mecánico	trasportatore *m* meccanico
tapis *m* convoyeur	cinta *f* transportadora	nastro *m* trasportatore
refroidir	enfriar	raffreddare
refroidissant *m*	refrigerante *m*	refrigerante *m*
capacité *f* de refroidissement	capacidad *f* de refrigeración	capacità *f* di raffreddamento
canal *m* de refroidissement	canal *m* de refrigeración	canale *m* di raffreddamento
filière *f* froide	boquilla *f* fría	filiera *f* raffreddata
enveloppe *f* de refroidissement	camisa *f* de refrigeración	camicia *f* di raffreddamento
mélangeur *m* réfrigérant	mezclador *m* refrigerante	mescolatore *m* raffreddatore
vitesse *f* de refroidissement	velocidad *f* de enfriamiento	velocità *f* di raffreddamento
anneau *m* de refroidissement	anillo *m* de enfriamiento	anello *m* di raffreddamento
temps *m* de refroidissement	tiempo *m* de enfriamiento	tempo *m* di raffreddamento

Nr.	English	Deutsch
308	**cooling water circuit**	Kühlwasserkreislauf *m*
309	**copolymer**	Copolymer *n*
310	**copolymerisation**	Copolymerisation *f*
311	**copy milling machine**	Kopierfräsmaschine *f*
312	**core displacement**	Kernversatz *m*
313	**core insert**	Kerneinsatz *m*
314	**core pin**	Kernlochstift *m*
315	**core puller**	Kernzug *m*
316	**corona treatment**	Coronabehandlung *f*
317	**co-rotating**	gleichläufig
318	**corrosion**	Korrosion *f*
319	**corrosion resistance**	Korrosionsbeständigkeit *f*
320	**corrugated panel**	Wellplatte *f*
321	**corrugated pipe**	Wellrohr *n*
322	**counter-pressure process**	Gegendruckverfahren *n*

Français	*Español*	*Italiano*
circuit *m* de refroidissement	circuito *m* del agua de refrigeración	circuito *m* di raffreddamento ad acqua
copolymère *m*	copolímero *m*	copolimero *m*
copolymérisation *f*	copolimerización *f*	copolimerizzazione *f*
machine *f* à copier	fresadora *f* copiadora	fresatrice *f* a copiare
déplacement *m* du noyau	desplazamiento *m* del núcleo	spostamento *m* del nucleo
insert *m* de noyau	inserto *m* macho	inserto *m* all'interno della cavità
broche *f* de noyau	espiga *f* del núcleo	spinotto *m*
tire-noyau *m*	núcleo *m* de tracción	nucleo *m* di trazione
traitement *m* corona	tratamiento *m* corona	trattamento *m* corona
co-rotatif	corrotativo	corotante
corrosion *f*	corrosión *f*	corrosione *f*
résistance *f* à la corrosion	resistencia *f* a la corrosión	resistenza *f* alla corrosione
plaque *f* ondulée	placa *f* ondulada	pannello *m* ondulato
tuyau *m* annelé	tubo *m* ondulado	tubo *m* ondulato
procédé *m* en contre-pression	procedimiento *m* de contrapresión	processo *m* contropressione

Nr.	English	Deutsch
323	**counter-rotating**	gegenläufig
324	**coupling agent**	Haftvermittler *m*
325	**crack**	Riss *m*
326	**crack formation**	Rissbildung *f*
327	**craze**	Craze *m;* Mikroriss *m*
328	**crazing**	Mikrorissbildung *f*
329	**creep**	Kriechen *n*
330	**creep curve**	Kriechkurve *f*
331	**creep modulus**	Kriechmodul *m*
332	**creep strength**	Zeitstandfestigkeit *f*
333	**creep test**	Zeitstandversuch *m*
334	**cresol resin**	Kresolharz *n*
335	**cross cutter**	Querschneider *m*

Français	*Español*	*Italiano*
contra-rotatif	giro en sentido contrario	controrotanti
agent *m* de couplage	agente *m* de adherencia	agente *m* di accoppiamento
fissure *f*	grieta *f;* fisura *f*	fessura *f*
formation *f* de fissures	formación *f* de grietas; fisuración *f*	fessurazione *f*
micro fendillement *m*	microfisura *f*	microfessura *f*
formation *f* de micro fendillement	formación *f* de microfisuras; cuarteamiento *m*	microfessurazione *f*
fluage *m*	fluencia *f*	scorrimento *m*
courbe *f* de fluage	curva *f* de fluencia	curva *f* di scorrimento
module *m* de fluage	módulo *m* de fluencia	modulo *m* di scorrimento
résistance *f* au fluage	resistencia *f* a la fluencia	resistenza *f* allo scorrimento
test *m* de fluage	ensayo *m* de fluencia	prova *f* di scorrimento
résine *f* crésol	resina *f* cresólica	resina *f* creosolo
découpeur *m* transversal	dispositivo *m* de corte transversal	dispositivo *m* di taglio trasversale

Nr.	English	Deutsch
336	cross-linking	Vernetzung *f*
337	cross-linking agent	Vernetzungsmittel *n*
338	cross-section	Querschnitt *m*
339	crosshead die	Umlenkkopf *m*
340	crystallinity	Kristallinität *f*
341	crystallisation	Kristallisation *f*
342	cure	Aushärtung *f*
343	cured	ausgehärtet
344	curing	Härtung *f*
345	curing agent	Härtemittel *n*
346	curing temperature	Aushärtetemperatur *f*
347	curing time	Härtezeit *f*
348	curling	Rollneigung *f*
349	custom moulder	Lohnspritzer *m*
350	cut	Zuschnitt *m*

Français	Español	Italiano
réticulation *f*	reticulación *f*; entrecruzamiento *m*	reticolazione *f*
agent *m* de réticulation	agente *m* reticulante	agente *m* di reticolazione
section *f* transversale	sección *f* transversal	sezione *f* trasversale
filière *f* en équerre	cabezal *m* transversal	filiera *f* con testa a croce
cristallinité *f*	cristalinidad *f*	cristallinità *f*
cristallisation *f*	cristalización *f*	cristallizzazione *f*
cuisson *f*	endurecimiento *m*	polimerizzazione *f*
réticulé	curado; reticulado	reticolato
cuisson *f*; réticulation *f*	curación *f*; reticulación *f*	reticolazione *f*
agent *m* de réticulation	agente *m* de curado	agente *m* vulcanizzante
température *f* de cuisson	temperatura *f* de curado	temperatura *f* di polimerizzazione
temps *m* de cuisson	tiempo *m* de curado	tempo *m* di polimerizzazione
ondulation *f* du rouleau	tendencia *f* al enrollamiento	arricciamento *m*
transformateur *m* sous-traitant	transformador *m* bajo pedido	stampatore *m* per conto terzi
flan *m*	corte *m*	taglio *m*

Nr.	English	Deutsch
351	**cutting device**	Schneidevorrichtung *f*
352	**cutting edge**	Schneidkante *f*
353	**cutting machine**	Schneidemaschine *f*
354	**cutting tool**	Schneidwerkzeug *n*
355	**cutting width**	Schnittbreite *f*
356	**cycle**	Zyklus *m*
357	**cycle time**	Zykluszeit *f*
358	**cylinder**	Zylinder *m*
359	**cylinder head**	Zylinderkopf *m*
360	**cylinder heater**	Zylinderheizung *f*
361	**dart-drop-test**	Fallbolzenprüfung *f*; Dart-drop-Test *m*
362	**daylight**	Werkzeugeinbauhöhe *f*
363	**dead spots**	Totecken *fpl*
364	**decomposition temperature**	Zersetzungstemperatur *f*
365	**decompression zone**	Dekompressionszone *f*

Français	*Español*	*Italiano*
unité *f* de découpe	dispositivo *m* de corte	dispositivo *m* di taglio
bord *m* coupant	canto *m* de corte	spigolo *m* di taglio
machine *f* de découpe	máquina *f* de corte	taglierina *f*
outil *m* de coupe	herramienta *f* de corte; cuchilla *f*	utensile *m* di taglio
largeur *f* de coupe	ancho *m* de corte	ampiezza *f* di taglio
cycle *m*	ciclo *m*	ciclo *m*
temps *m* de cycle	tiempo *m* de ciclo	tempo *m* di ciclo
cylindre *m*	cilindro *m*	cilindro *m*
tête *f* de cylindre	cabezal *m* del cilindro	testa *f* del cilindro
réchauffeur *m* de cylindre	calefactor *m* del cilindro	riscaldatore *m* del cilindro
test *m* de la chute de bille (dart)	ensayo *m* de caída del dardo	prova *f* di caduta del dardo
espace *m* entre les plateaux	apertura *f* entre platos	luce *f* tra i piani
taches *fpl* de mort	zonas *fpl* muertas	zone *fpl* morte
température *f* de décomposition	temperatura *f* de descomposición	temperatura *f* di decomposizione
zone *f* de décompression	zona *f* de descompresión	sezione *f* di decompressione

Nr.	English	Deutsch
366	**decorative laminates**	dekorative Schichtpressstoffe *mpl*
367	**deep drawing**	Tiefziehen *n*
368	**deflashing**	Entgraten *n*
369	**deflection**	Durchbiegung *f*
370	**deform, to**	verformen
371	**deformation**	Deformation *f;* Verformung *f*
372	**degassing**	Entgasen *n*
373	**degradable**	abbaubar
374	**degradable plastics**	abbaubare Kunststoffe *mpl*
375	**degradation**	Abbau *m*
376	**degree of crystallinity**	Kristallisationsgrad *m*
377	**degree of polymerisation**	Polymerisationsgrad *m*
378	**dehumidification**	Entfeuchtung *f*
379	**delamination**	Delaminierung *f*
380	**demould, to**	entformen
381	**density**	Dichte *f*

Français	Español	Italiano
panneaux *mpl* décoratifs	laminados *mpl* decorativos	laminati *mpl* decorativi
thermoformage *m* profond	conformado *m* al vacio	formatura *f* profonda
ébarbage *m;* ébavurage *m*	rebaba *f*	sbavatura *f*
déflexion *f*	deflección *f*	deflessione *f*
déformer	deformar	deformare
déformation *f*	deformación *f*	deformazione *f*
dégazage *m*	desgasificación *f*	degasaggio *m*
dégradable	degradable	degradabile
plastiques *mpl* dégradables	plásticos *mpl* degradables	materie *fpl* plastiche degradabili
dégradation *f*	degradación *f*	degradazione *f*
taux *m* de cristallinité	grado *m* de cristalización	grado *m* di cristallinità
taux *m* de polymérisation	grado *m* de polimerización	grado *m* di polimerizzazione
déshumidification *f*	deshumidificación *f*	deumidificazione *f*
délamination *f*	delaminación *f*	delaminazione *f*
démouler	desmoldear	sformare
densité *f*	densidad *f*	densità *f*

Nr.	English	Deutsch
382	**depolymerisation**	Depolymerisation *f*
383	**depth of draw**	Ziehtiefe *f*
384	**devolatilising**	Entgasen *n*
385	**devolatilising unit**	Entgasungseinheit *f*
386	**dew point**	Taupunkt *m*
387	**diaphragm gate**	Scheibenanguss *m*
388	**die**	Düse *f*
389	**die-face pelletiser**	Heißabschlaggranulator *m*
390	**die gap**	Düsenspalt *m*
391	**die immersion depth**	Düseneintauchtiefe *f*
392	**die land**	Bügellänge *f*
393	**die swell**	Düsenaufweitung *f*
394	**diecasting mould**	Druckgusswerkzeug *n*
395	**dielectric**	dielektrisch

Français	*Español*	*Italiano*
dépolymérisation *f*	depolimerización *f*	depolimerizzazione *f*
profondeur *f* de formage	profundidad *f* de moldeo	profondità *f* di stiro
dégazage *m*	desgasificación *f*	devolatilizzazione *f*
unité *f* de dégazage	unidad *f* de desgasificación	unità *f* di devolatilizzazione
point *m* de rosée	punto *m* de rocío	punto *m* di condensa
injection *f* en diaphragme	entrada *f* de diafragma; entrada *f* de disco	orifizio *m* di entrata a diaframma
filière *f*	boquilla *f*	filiera *f*
granulatrice *f* à chaud	granuladora *f* con corte en caliente	taglio *m* in testa
écartement *m* de filière	abertura *f* de la boquilla	apertura *f* della filiera
profondeur *f* d'immersion de la filière	longitud *f* de penetración de la boquilla	profondità *f* di immersione della filiera
lèvres *fpl* de la filière	labios *mpl* de la boquilla	labbro *m* della filiera
gonflement *m* de filière	dilatación *f* de la boquilla	rigonfiamento *m* nella filiera
moule *m* de coulée	molde *m* de colada	stampo *m* di colata
diélectrique	dieléctrico	dielèttrico

Nr.	English	Deutsch
396	**dielectric constant**	Dielektrizitätszahl *f*
397	**dielectric dissipation factor**	dielektrischer Verlustfaktor *m*
398	**dielectric strength**	Durchschlagfestigkeit *f*
399	**diffusion**	Diffusion *f*
400	**diisocyanate**	Diisocyanat *n*
401	**dilatency**	Dilatanz *f*
402	**diluent**	Verdünnungsmittel *n*
403	**dimension**	Abmessung *f*
404	**dimensional stability**	Maßbeständigkeit *f*; Maßhaltigkeit *f*
405	**dip coating**	Tauchlackierung *f*
406	**dip moulding**	Tauchformverfahren *n*
407	**dip, to**	tauchen
408	**direct gating**	Direktanspritzen *n*
409	**disc gate**	Scheibenanschnitt *m*
410	**discharge**	Austrag *m*

Français	*Español*	*Italiano*
constante *f* diélectrique	constante *f* dieléctrica	costante *f* dielèttrica
facteur *m* de dissipation diélectrique	factor *m* de disipación dieléctrica	fattore *m* di dissipazione dielèttrica
rigidité *f* diélectrique	rígidez *f* dieléctrica	rigidità *f* elettrica
diffusion *f*	difusión *f*	diffusione *f*
diisocyanate *m*	diisocianato *m*	diisocianato *m*
dilatance *f*	dilatancia *f*	dilatanza *f*
diluant *m*	diluyente *m*	diluente *m*
dimension *f*	dimensión *f*	dimensione *f*
stabilité *f* dimensionnelle	estabilidad *f* dimensional	stabilità *f* dimensionale
moulage *m* au trempé	recubrimiento *m* por inmersión	stampaggio *m* a immersione
moulage *m* au trempé	moldeo *m* por inmersión	stampaggio *m* a immersione
tremper	sumergir	immergere
injection *f* directe	inyección *f* directa	orifizio *m* di entrata diretta
injection *f* à disque	punto *m* de inyección de disco	punto *m* di iniezione a diaframma
décharge *f*	descarga *f*	scarico *m*

Nr.	English	Deutsch
411	**discharge device**	Austragsvorrichtung *f*
412	**discoloration**	Verfärbung *f*
413	**discoloured streaks**	Verfärbungsschlieren *fpl*
414	**disperse, to**	dispergieren
415	**dispersing agent**	Dispergierhilfsmittel *n*
416	**dispersion**	Dispersion *f*
417	**doctor knife**	Abstreifmesser *n;* Rakelmesser *n*
418	**dosing**	Dosieren *n*
419	**double toggle**	Doppelkniehebel *m*
420	**dough moulding compound (DMC)**	teigförmige Premix-Pressmasse *f*
421	**downstream equipment**	Nachfolgeeinrichtung *f*
422	**downtime**	Stillstandzeit *f*
423	**draft**	Konizität *f;* Entformungsschräge *f*
424	**drag flow**	Schleppströmung *f*
425	**draw ratio**	Reckverhältnis *n*

Français	*Español*	*Italiano*
unité *f* de décharge	dispositivo *m* de descarga	dispositivo *m* di scarico
décoloration *f*	decoloración *f*	scolorimento *m*
traces *fpl* de décoloration	vetas *fpl* decoloradas	segni *mpl* scoloriti
disperser	dispersar	disperdere
agent *m* dispersant	agente *m* dispersante	agente *m* di dispersione
dispersion *f*	dispersión *f*	dispersione *f*
racle *f*	cuchilla *f* de recubrimiento	raschiatore *m;* racla *f*
dosage *m*	dosificación *f*	dosaggio *m*
double genouillère *f*	doble rodillera *f*	ginocchiera *f* doppia
DMC *m*	compuesto *m* para moldeo espesado físicamente (DMC)	composto *m* per stampaggio di masse
équipement *m* aval	equipo *m* de tratamiento secundario	attrezzatura *f* a valle
temps *m* mort	tiempo *m* muerto	tempo *m* passivo
angle *m* de démoulage	conicidad *f;* ángulo *m* de desmoldeo	conicità *f*
flux *m* tangentiel	flujo *m* de avance; flujo *m* principal	flusso *m* totale
taux *m* d'étirage	relación *f* de estiraje	rapporto *m* di stiro

Nr.	English	Deutsch
426	**draw-warping process**	Kettstreckverfahren *n*
427	**drawdown**	Auslängung *f*
428	**drilling**	Bohren *n*
429	**drive**	Antrieb *m*
430	**drive shaft**	Antriebswelle *f*
431	**drive unit**	Antrieb *m*
432	**drop height**	Fallhöhe *f*
433	**drop test**	Fallversuch *m*
434	**drum film casting**	Trommelgießverfahren *n*
435	**dry-air drier**	Trockenlufttrockner *m*
436	**dry blend**	Dryblend *n*
437	**dry colouring**	Trockeneinfärben *n*
438	**dry cycle time**	Trockenlaufzeit *f;* Trockenzyklus *m*
439	**dry residue**	Trockenrückstand *m*

Français	Español	Italiano
procédé *m* d'étirage-déformation	proceso *m* de estirado de la urdimbre	processo *m* di stiro-deformazione
allongement *m*	descuelgue *m*	riduzione *f* per stiro
perçage *m*	taladrado *m*	perforazione *f*
moteur *m*	accionamiento *m*	comando *m*
entraînement *m*	eje *m* del accionamiento	albero *m* di comando
unité *f* d'entraîne-ment	accionamiento *m*	unità *f* di comando
hauteur *f* de chute	altura *f* de caída	altezza *f* di caduta
test *m* de chute	ensayo *m* de caída	prova *f* d'urto per caduta
coulée *f* de film au tambour	colada *f* de película por tambor	colata *f* di film con tamburo
sécheur *m* à air sec	secador *m* de aire seco	essiccatore *m* ad aria secca
mélange *m* à sec; dryblend *m*	mezcla *f* seca	miscela *f* secca
coloration *f* à sec	coloración *f* en seco; coloración *f* en la masa	coloritura *f* a secco
temps *m* de cycle à vide	tiempo *m* de funcionamiento en seco	tempo *m* di ciclo a secco
extrait *m* sec	residuo *m* seco	residuo *m* secco

Nr.	English	Deutsch
440	**dryer**	Trockner *m*
441	**drying**	Trocknen *n*
442	**drying temperature**	Trocknungstemperatur *f*
443	**drying time**	Trocknungszeit *f*
444	**ductile fracture**	Zähbruch *m*
445	**dwell time**	Verweilzeit *f*
446	**dye**	Farbstoff *m*
447	**easy flowing**	leichtfließend
448	**edge gate**	seitlicher Anguss *m*
449	**edge trim**	Randbeschnitt *m*
450	**efflorescence**	Ausblühen *n*
451	**ejection**	Auswerfen *n*
452	**ejector**	Auswerfer *m*
453	**ejector bolt**	Auswerferbolzen *m*
454	**ejector bush**	Auswerferhülse *f*
455	**ejector pin**	Auswerferstift *m*

Français	*Español*	*Italiano*
sécheur *m*	secador *m*	essiccatore *m*
séchage *m*	secado *m*	essiccamento *m*
température *f* de séchage	temperatura *f* de secado	temperatura *f* di essiccazione
temps *m* de séchage	tiempo *m* de secado	tempo *m* di essiccazione
rupture *f* ductile	fractura *f* dúctil	frattura *f* duttile
temps *m* de mélange	pausa *f* de cierre	tempo *m* di attesa
colorant *m*	tinte *m*	colorante *m*
fluide	de fácil flujo	scorrimento *m* veloce
injection *f* en coin	entrada *f* lateral	iniezione *f* a filo
ébarbage *m*	corte *m* de los bordes	cimosa *f*
efflorescence *f*	eflorescencia *f*	efflorescenza *f*
éjection *f*	expulsión *f*	espulsione *f*
éjecteur *m*	expulsor *m*	estrattore *m*
extracteur *m*	bulón *m* de expulsión	bullone *m* estrattore
éjection *f* tubulaire	manguito *m* del expulsor	boccola *f* dell'estrattore
aiguille *f* d'éjection	espiga *f* de expulsión	perno *m* dell'estrattore

Nr.	English	Deutsch
456	**ejector plate**	Auswerferplatte *f*
457	**ejector ring**	Auswerferring *m*
458	**ejector rod**	Auswerferstange *f*
459	**elasticity**	Elastizität *f*
460	**elastomer**	Elastomer *n*
461	**elbow**	Krümmer *m*
462	**electric discharge machining**	Funkenerosion *f*
463	**electrical conductivity**	elektrische Leitfähigkeit *f*
464	**electrodeposition coating**	Elektrotauchlackierung *f*
465	**electroluminescence**	Elektrolumineszenz *f*
466	**electrolyte**	Elektrolyt *m*
467	**electromagnetic shielding**	elektromagnetische Abschirmung *f*
468	**electroplating**	Galvanisieren *n*
469	**electrostatic charge**	elektrostatische Aufladung *f*

Français	*Español*	*Italiano*
plateau *m* d'éjection	placa *f* expulsora	piastra *f* di estrazione
anneau *m* d'éjection	anillo *m* expulsor	anello *m* dell'estrattore
tige *f* d'éjection	varilla *f* expulsora	barra *f* dell'estrattore
élasticité *f*	elasticidad *f*	elasticità *f*
élastomère *m*	elastómero *m*	elastomero *m*
coude *m*	codo *m*	raccordo *m* a gomito
électroérosion *f*	electroerosión *f*	elettroerosione *f*
conductivité *f* électrique	conductividad *f* eléctrica	conduttività *f* elettrica
revêtement *m* par électrodéposition	revestimiento *m* por electrodeposición	rivestimento *m* per elettrodeposizione
électroluminescence *f*	electroluminiscencia *f*	elettroluminescenza *f*
électrolyte *m*	electrólito *m*	elettrolito *m*
protection *f* contre les courants électromagnétiques	escudo *m* electromagnético	schermatura *f* elettromagnetica
galvanisation *f*	galvanización *f*	galvanoplastica *f*
charge *f* électrostatique	carga *f* electrostática	carica *f* elettrostatica

Nr.	English	Deutsch
470	**electrostatic powder coating**	elektrostatisches Pulversprühverfahren *n*
471	**elongation**	Dehnung *f*
472	**elongation at break**	Bruchdehnung *f;* Reißdehnung *f*
473	**embedding**	Einbetten *n*
474	**embossing**	Prägen *n;* Gaufrieren *n*
475	**embossing calender**	Prägekalander *m*
476	**embossing roller**	Prägewalze *f*
477	**EMI shielding**	elektromagnetische Abschirmung *f*
478	**emulsifier**	Emulgator *m*
479	**emulsion**	Emulsion *f*
480	**emulsion polymerisation**	Emulsionspolymerisation *f*
481	**encapsulating**	Einkapseln *n*
482	**engineering plastics**	Konstruktionskunststoffe *mpl;* Technische Kunststoffe *mpl*
483	**engraving**	Gravieren *n*

Français	*Español*	*Italiano*
revêtement *m* de poudre électrostatique	recubrimiento *m* por polvo electrostático	rivestimento *m* con polvere elettrostatica
allongement *m*	alargamiento *m*	allungamento *m*
allongement *m* à la rupture	alargamiento *m* a la rotura	allungamento *m* a rottura
enrobage *m*	inclusión *m*	inclusione *m*
gaufrage *m*	estampado *m;* gofrado *m*	goffratura *f*
calandre *f* de gaufrage	calandra *f* de gofrado	calandra *f* di goffratura
cylindre *m* de gaufrage	cilindro *m* gofrador	cilindro *m* di goffratura
protection *f* contre les interférences électromagnétiques	protección *f* electromagnética	schermatura *f* elettromagnetica
émulsifiant *m*	emulsionante *m*	emulsionante *m*
émulsion *f*	emulsión *f*	emulsione *f*
polymérisation *f* en émulsion	polimerización *f* en emulsión	polimerizzazione *f* per emulsione
encapsulation *f*	encapsulado *m*	incapsulamento *m*
plastiques *mpl* techniques	plásticos *mpl* técnicos	tecnopolimeri *mpl*
gravure *f*	grabado *m*	incisione *f*

Nr.	English	Deutsch
484	**entrapped air**	Lufteinschluss *m*
485	**environmental stress cracking**	umgebungsbedingte Spannungsrisskorrosion *f*
486	**epoxidation**	Epoxidation *f*
487	**epoxy resin** *(EP)*	Epoxidharz *n*
488	**equilibrium**	Gleichgewicht *n*
489	**equipment** *(machine)*	Ausrüstung *f*
490	**etch, to**	ätzen
491	**ethylene**	Ethylen *n*
492	**ethylene-propylene rubber**	Ethylen-Propylen-Kautschuk *m*
493	**ethylene vinyl acetate copolymer** *(EVA)*	Ethylen-Vinylacetat-Copolymer *n*
494	**evaporation**	Verdampfung *f*
495	**exothermic**	exotherm
496	**expandable**	expandierbar
497	**expandable polystyrene**	schäumbares Polystyrol *n*
498	**expander roll**	Breithaltewalze *f*

Français	Español	Italiano
inclusion *f* d'air	inclusión *f* de aire	inclusione *f* di aria
fissuration *f* à l'environnement (aux intempéries)	rotura *f* por tensión ante degradación medioambiental	fessurazione *f* da sollecitazione ambientale
époxydation *f*	epoxidación *f*	epossidazione *f*
résine *f* époxyde	resina *f* epoxídica	resina *f* epossidica
équilibre *m*	equilibrio *m*	equilibrio *m*
équipement *m*	equipo *m;* utillaje *m*	attrezzatura *f;* apparecchiatura *f*
corroder	corroer	corrodere
éthylène *m*	etileno *m*	etilene *m*
caoutchouc *m* éthylène-propylène	caucho *m* de etileno-propileno	elastomero *m* etilene propilene
copolymère *m* d'éthylène-acétate de vinyle	copolímero *m* de etileno-acetato de vinilo	copolimero *m* etilene acetato di vinile
évaporation *f*	evaporización *f*	evaporazione *f*
exothermique	exotermo	esoterma *f*
expansible	expandible	espandibile
polystyrène *m* expansible	poliestireno *m* expandible	polistirene *m* espandibile
rouleau *m* anti plis	rodillo *m* de expansión	rullo *m* di espansione

Nr.	English	Deutsch
499	**extender**	Streckungsmittel *n*
500	**extensibility**	Dehnbarkeit *f*
501	**external calibration**	Außenkalibrierung *f*
502	**extruder**	Extruder *m*
503	**extruder head**	Extruderkopf *m*
504	**extrusion**	Extrusion *f*
505	**extrusion blow moulding**	Extrusionsblasformen *n*
506	**extrusion coating**	Extrusionsbeschichten *n*
507	**extrusion die**	Extrusionswerkzeug *n*
508	**extrusion line**	Extrusionsanlage *f*
509	**extrusion welding**	Extrusionsschweißen *n*
510	**exudation**	Ausschwitzen *n*
511	**fabric**	Gewebe *n*
512	**failure**	Versagen *n*
513	**failure analysis**	Versagensanalyse *f*

Français	*Español*	*Italiano*
diluant *m*	extendedor *m*	estensore *m*
extensibilité *f*	extensibilidad *f*	estensibilità *f*
calibrage *m* externe	calibración *f* externa	calibrazione *f* esterna
extrudeuse *f*	extrusora *f*	estrusore *m*
tête *f* d'extrusion	cabezal *m* de extrusión	testa *f* d'estrusione
extrusion *f*	extrusión *f*	estrusione *f*
extrusion-soufflage *f*	moldeo *m* por extrusión-soplado	estrusione *f* soffiaggio
extrusion-couchage *f*	recubrimiento *m* por extrusión	rivestimento *m* per estrusione
filière *f*	boquilla *f* de extrusión	filiera *f* di estrusione
ligne *f* d'extrusion	línea *f* de extrusión	linea *f* di estrusione
soudage *m* par extrusion	soldadura *f* por extrusión	saldatura *f* per estrusione
exsudation *f*	exudación *f*	essudazione *f*
tissu *m*	tejido *m*	tessuto *m*
défaut *m*	defecto *m*, rechazo *m*	cedimento *m*
analyse *f* des défauts	análisis *m* de defectos	analisi *f* del cedimento

Nr.	English	Deutsch
514	**failure criteria**	Versagenskriterien *npl*
515	**falling ball test**	Kugelfallversuch *m*
516	**falling dart test**	Fallbolzentest *m*
517	**family mould; multi-cavity mould**	Mehrfach-Werkzeug *n*
518	**fan gate**	Fächeranguss *m*
519	**fastener**	Verbindungselement *n*
520	**fatigue**	Ermüdung *f*
521	**fatigue fracture**	Ermüdungsbruch *m*
522	**fatigue resistance**	Ermüdungswiderstand *m*
523	**fatigue test**	Dauerschwingversuch *m*
524	**feed**	Materialzufuhr *f*
525	**feed hopper**	Einfülltrichter *m;* Aufgabetrichter *m*
526	**feed plate**	Anguss-Verteilerplatte *f*
527	**feed pocket; feeding throat**	Einzugstasche *f*

Français	*Español*	*Italiano*
critères *mpl* de défaut	criterios *mpl* del fallo	criteri *mpl* di cedimento
essai *m* à la bille	ensayo *m* de caída de la bola	prova *f* di caduta della sfera
test *m* «dart»	ensayo *m* de caída del dardo	prova *f* di caduta del dardo
moule *m* multi-empreinte	molde *m* apilado	stampo *m* famiglia
injection *f* en éventail	entrada *f* en forma de abanico	entrata *f* a ventaglio
agent *m* de fixation	elemento *m* de fijación	dispositivo *m* di fissaggio
fatigue *f*	fatiga *m*	fatica *f*
rupture *f* à la fatigue	fractura *f* a la fatiga	frattura *f* da fatica
résistance *f* à la fatigue	resistencia *f* a la fatiga	resistenza *f* alla fatica
test *m* de fatigue	ensayo *m* de fatiga	prova *f* di fatica
alimentation *f*	alimentación *f*	alimentazione *f*
trémie *f* d'alimentation	tolva *f* de alimentación	tramoggia *f* di carico
plaque *f* d'alimentation	placa *f* del distribuidor	piastra *f* di alimentazione
goulotte *f* d'alimentation	abertura *f* de alimentación	tasca *f* di alimentazione; canale *m* di alimentazione

Nr.	English	Deutsch
528	**feed point**	Anspritzpunkt *m*
529	**feed pump**	Förderpumpe *f*
530	**feed ram**	Dosierkolben *m*
531	**feed roll**	Einzugswalze *f*
532	**feed section**	Einzugszone *f;* Förderlänge *f*
533	**feed, to**	dosieren
534	**feeder**	Dosiervorrichtung *f*
535	**feeding aid**	Einzugshilfe *f*
536	**feeding hopper**	Fülltrichter *m*
537	**feeding stroke**	Dosierweg *m*
538	**female mould**	Gesenk *n;* Matrize *f*
539	**fibre**	Faser *f*
540	**fibre composite**	Faserverbundwerkstoff *m*

Français	*Español*	*Italiano*
point *m* d'injection	entrada *f* puntiforme	punto *m* di alimentazione
pompe *f* doseuse	bomba *f* de dosificación	pompa *f* di alimentazione; pompa *f* dosatrice
piston *m* d'alimentation	pistón *m* de alimentación	pistone *m* di alimentazione
cylindre *m* d'alimentation	cilindro *m* de alimentación	cilindro *m* di alimentazione
zone *f* d'alimentation	zona *f* de entrada; zona *f* de alimentación	zona *f* di alimentazione
doser	dosificar	dosare; alimentare
dispositif *m* de dosage	dispositivo *m* de dosificación	dosatore *m*
agent *m* d'alimentation	alimentación *f* asistida	agente *m* di alimentazione
trémie *f* d'alimentation	tolva *f* de alimentación	tramoggia *f* di alimentazione
course *f* d'alimentation	recorrido *m* de la alimentación	corsa *f* di alimentazione
matrice *f*	placa *f* de la cavidad; matriz *f*	stampo *m* femmina
fibre *f*	fibra *f*	fibra *f*
composite *m* à base de fibres	composite *m* a base de fibra	composito *m* con fibre

Nr.	English	Deutsch
541	**fibre reinforcement**	Faserverstärkung *f*
542	**filament**	Filament *n*
543	**filament winding**	Wickelverfahren *n*
544	**filler**	Füllstoff *m*
545	**filling index**	Füllindex *m*
546	**filling pattern method**	Füllbildmethode *f*
547	**filling stage**	Füllvorgang *m*
548	**film** *(plastic)*	Folie *f;* Film *m*
549	**film blowing**	Folienblasen *n*
550	**film bubble**	Folienschlauch *m*
551	**film casting**	Filmgießen *n*
552	**film gate**	Filmanguss *m*
553	**film haul-off**	Folienabzug *m*
554	**film stretching plant**	Folienreckanlage *f*

Français	*Español*	*Italiano*
fibre *f* de renforcement	refuerzo *m* de fibra	rinforzo *m* con fibre
filament *m*	filamento *m*	filamento *m*
enroulement *m* filamentaire	enrollado *m* de filamentos	avvolgimento *m* filamentare
charge *f*	carga *f*	carica *f*
taux *m* de remplissage	índice *m* de llenado	indice *m* di riempimento
méthode *f* de remplissage d'image	método *m* de imagen de llenado	metodo *m* per riempimento di immagini
étape *f* de remplissage	proceso *m* de carga	stadio *m* di riempimento
film *m*	película *f*; hoja *f*	film *m*; pellicola *f*
soufflage *m* de gaine	soplado *m* de películas	filmatura *f* per soffiaggio
bulle *f* de film soufflé	manga *f* de película soplada	bolla *f* di film
coulée *f* de film	colada *f* de película	filmatura *f* per colata
injection *f* en voile	entrada *f* pelicular	entrata *f* piatta
tirage *m* de film	arrastre *m* de la película	traino *m* di film
ligne *f* d'étirage des feuilles	instalación *f* de película estirada	impianto *m* di stiro di film

Nr.	English	Deutsch
555	**film tape**	Folienbändchen *n*
556	**film web**	Folienbahn *f*
557	**film winder**	Folienwickler *m*
558	**finishing**	Nachbearbeitung *f*
559	**fire behaviour**	Brandverhalten *n*
560	**fish eye**	Stippe *f*
561	**fish tail die**	Fischschwanzdüse *f*
562	**fit**	Passung *f*
563	**fixed platen**	Werkzeugaufspannplatte *f*, feststehende; Düsenplatte *f*
564	**flame bonding**	Flammkaschieren *n*
565	**flame polishing**	Flammpolieren *n*
566	**flame resistance**	Schwerentflammbarkeit *f;* Flammwidrigkeit *f*
567	**flame retardant**	Flammschutzmittel *n*
568	**flame-spraying**	Flammspritzen *n*

Français	*Español*	*Italiano*
ruban *m*	cintas *fpl* de hojas; rafia *f*	nastro *m*
bande *f* de film	película *f* continua	nastro *m* di film
enrouleur *m*	bobinador *m*	avvolgitore *m*
finition *f*	acabado *m;* desbarbado *m*	finitura *f*
comportement *m* au feu	comportamiento *m* al fuego	comportamento *m* al fuoco
peau *f* de requin	ojo *m* de pez	infuso *m*
filière *f* en queue de poisson	boquilla *f* en "cola de pescado"	filiera *f* a coda di pesce
ajustage *m*	ajuste *m*	tolleranza *f*
plateau *m* fixe du moule	placa *f* fija del molde	piastra *f* d'impronta fissa
adhésion *f* à la flamme	recubrimiento *m* a la llama	accoppiamento *m* alla fiamma
polissage *m* à la flamme	pulido *m* a la llama	finitura *f* alla fiamma
résistance *f* à la flamme	resistencia *f* al fuego	resistenza *f* al fuoco
ignifugeant *m;* retardateur *m* de flamme	retardador *m* de llama	ritardante *m* di fiamma
flammage *m*	proyección *f* a la llama	fiammatura *f*

Nr.	English	Deutsch
569	flame treatment	Beflammen *n*
570	flammability	Entflammbarkeit *f*
571	flange	Flansch *m*
572	flash	Grat *m;* Butzen *m*
573	flash formation	Gratbildung *f*
574	flash mould	Abquetschwerkzeug *n*
575	flash point	Flammpunkt *m*
576	flash ridge	Abquetschrand *m*
577	flat sheet film	Flachfolie *f*
578	flex-lip die	Flex-Lip-Düse *f*
579	flexible bag moulding	Gummisack-Verfahren *n*
580	flexographic printing	Flexodruck *m*
581	flexural modulus	Biegemodul *m*
582	flexural strength	Biegefestigkeit *f*

Français	*Español*	*Italiano*
traitement *m* à la flamme	tratamiento *m* con llama	trattamento *m* alla fiamma
inflammabilité *f*	inflamabilidad *f*	infiammabilità *f*
bride *f*	brida *f*	flangia *f*
bavure *f*	rebaba *f*	bava *f*
formation *f* de bavures	formación *f* de rebaba	formazione *f* di bava
moule *m* à échappement	molde *m* de rebaba	stampo *m* con linea di bava
point *m* d'inflammation spontanée; flash point	punto *m* de inflamación	punto *m* di fiamma
bord *m* d'appui	borde *m* de rebaba; borde *m* de aplastamiento	sfogo *m*
feuille *f* à plat	lámina *f* plana; hoja *f* plana	foglia *f* piana
filière *f* à lèvre souple	hilera *f* de labios flexibles	filiera *f* a labbro (barra) flessibile
moulage *m* au sac	moldeo *m* con bolsa de goma	stampaggio *m* con membrana
flexographie *f*	flexografía *f*	flessografia *f*
module *m* de flexion	módulo *m* de flexión	modulo *m* a flessione
résistance *f* à la flexion	resistencia *f* a la flexión	resistenza *f* alla flessione

Nr.	English	Deutsch
583	**flexural stress**	Biegespannung *f*
584	**flight** *(of a screw)*	Gang *m*; Schneckengang *m*
585	**flight depth**	Gangtiefe *f*
586	**flight depth ratio**	Gangtiefenverhältnis *n*
587	**flight hard-facing**	Stegpanzerung *f*
588	**flight land clearance**	Schneckenspalt *m*
589	**flight land width**	Schneckenstegbreite *f*
590	**flocking**	Beflocken *n*
591	**flow agent**	Fließhilfe *f*
592	**flow birefringence**	Strömungsdoppelbrechung *f*
593	**flow control valve**	Stromregelventil *n*
594	**flow curve**	Fließkurve *f*
595	**flow line**	Bindenaht *f*
596	**flow path**	Fließweg *m*

Français	*Español*	*Italiano*
contrainte *f* en flexion	esfuerzo *m* de flexión	sollecitazione *f* alla flessione
filet *m*	hilo *m;* filete *m*	filetto *m*
profondeur *f* de filetage	profundidad *f* de filete	profondità *f* del filetto
profondeur *m* de filet	relación *m* de la altura del filete	rapporto *m* di profondità del filetto
blindage *m* de filet	templado *m* del filete	temperatura *f* del filetto
pas *m* de vis	holgura *f* del tornillo	gioco *m* del filetto
largeur *m* du filet	anchura *f* del filete	spessore *m* del filetto
flocage *m*	flocado *m*	floccaggio *m*
fluidifiant *m*	agente *m* de fluidez	agente *m* fluidificante
biréfringence *f* du flux	flujo *m* birrefringente	birifrangenza *f* del flusso
valve *f* de régulation de flux	válvula *f* de regulación	valvola *f* di controllo del flusso
courbe *f* de flux	curva *f* de flujo	curva *f* del flusso
ligne *f* de flux	linea *f* de reunión de flujos	linea *f* di flusso
chemin *m* de l'écoulement	recorrido *m* del flujo	canale *m* di adduzione

Nr.	English	Deutsch
597	**flow rate**	Durchflussgeschwindigkeit *f*
598	**flow resistance**	Fließwiderstand *m*
599	**flowability**	Fließfähigkeit *f*
600	**fluid mixer**	Fluidmischer *m*
601	**fluidised bed coating**	Wirbelsintern *n*
602	**fluorescence**	Fluoreszenz *f*
603	**fluorination**	Fluorierung *f*
604	**fluoropolymer**	Fluorpolymer *n*
605	**foam**	Schaumstoff *m*
606	**foam moulding**	Formschäumen *n*
607	**foaming**	Schäumen *n*
608	**foaming agent**	Schäummittel *n*
609	**foaming machine**	Schäumanlage *f*
610	**fogging**	Fogging *n*
611	**foil** *(metal)*	Folie *f*

Français	*Español*	*Italiano*
vitesse *f* d'écoulement	velocidad *f* de flujo	indice *m* di fluidità
résistance *f* de la matière à l'état fondu	resistencia *f* al flujo	resistenza *f* allo scorrimento
coulabilité *f*	fluidez *f*	fluidificazione *f*
mélangeur *m* liquide	mezclador *m* de fluidos	mescolatore *m* per liquidi
revêtement *m* en lit fluidisé	recubrimiento *m* en lecho fluidificado	rivestimento *m* in letto fluidizzato
fluorescence *f*	fluorescencia *f*	fluorescenza *f*
fluoration *f*	fluorado *m*	fluorazione *f*
polymère *m* fluoré	polímero *m* flurocarbonado	polimero *m* fluorurato
mousse *f*	espuma *f*	schiuma *f*; espanso *m*
moulage *f* de mousse	moldeo *m* de espuma	stampaggio *m* di espanso
moussage *m*	espumación *f*	espansione *f*
agent *m* de moussage; agent *m* d'expansion	agente *m* de expansión	agente *m* espandente
machine *f* à mousser	instalación *f* de espumación	macchina *f* per espansi
fogging *m*	fogging *m*	appannamento *m*
feuille *f*	hoja *f*	foglio *m*

Nr.	English	Deutsch
612	**folding**	Abkanten *n*
613	**form-fill-seal machine (FFS)**	Form-Füll-Siegelmaschine *f* (FFS)
614	**forming**	Umformen *n*
615	**fracture**	Bruch *m*
616	**fracture behaviour**	Bruchverhalten *n*
617	**fracture mechanics**	Bruchmechanik *f*
618	**fresh-air drier**	Frischlufttrockner *m*
619	**friction**	Reibung *f*
620	**friction welding**	Reibschweißen *n*
621	**frictional heat**	Friktionswärme *f*
622	**frictional resistance**	Reibungswiderstand *m*
623	**frost line**	Frostlinie *f*
624	**furan resins**	Furanharze *npl*

Français	*Español*	*Italiano*
pliage *m*	doblado *m*	piegatura *f*
machine FFS *f* (formage remplissage scellage)	máquina *f* de moldeo, llenado y sellado	macchina *f* formatrice riempitrice sigillatrice
formage *m*	conformado *m*	formatura *f*
rupture *f*	fractura *f*	frattura *f*
comportement *m* à la rupture	comportamiento *m* de fractura	comportamento *m* alla rottura
mécanique *f* de rupture	fractura *f* mecánica	meccanica *f* della rottura
sécheur *m* à air froid	secador *m* de aire fresco	essiccatore *m* ad aria ambiente
frottement *m*	fricción *f*	frizione *f*
soudage *m* par friction	soldadura *f* por fricción	saldatura *f* per frizione
chaleur *f* de frottement	calor *m* de fricción	calore *m* di frizione
résistance *f* au frottement (à la friction)	resistencia *f* a la fricción	resistenza *f* a frizione
ligne *f* de solidification	línea *f* de congelación	linea *f* di raffreddamento
résines *fpl* furanniques	resinas *fpl* furánicas	resine *fpl* furaniche

Nr.	English	Deutsch
625	**fusible core technique**	Schmelzkerntechnik *f*
626	**gas** *(assisted)* **injection moulding**	Gas-Innendruckverfahren *n;* Gas-Innendruck-Spritzgießverfahren *n*
627	**gas-counter-pressure process**	Gas-Gegendruckverfahren *n*
628	**gas injection extruder**	Begasungsextruder *m*
629	**gas permeability**	Gasdurchlässigkeit *f*
630	**gas-phase polymerisation**	Gasphasenpolymerisation *f*
631	**gaseous**	gasförmig
632	**gas-nitrided**	gasnitriert
633	**gate**	Anschnitt *m;* Anguss *m*
634	**gear pump**	Zahnradpumpe *f*
635	**gear rack**	Zahnstange *f*
636	**gear wheel**	Zahnrad *n*
637	**gel**	Gel *n*

Français	Español	Italiano
technique *f* à noyaux fusibles	técnica *f* de noyo fundible	tecnica *f* del nucleo fusibile
moulage *f* sous pression interne de gaz	moldeo *m* por inyección asistido por gas	stampaggio *m* con gas
procédé *m* à contre-pression de gaz	proceso *m* a contrapresión de gas	processo *m* a contropressione di gas
extrusion *f* à injection de gaz	extrusora *f* de gasificación directa	estrusore *m* a iniezione con gas
perméabilité *f* au gaz	permeabilidad *f* a los gases	permeabilità *f* ai gas
polymérisation *f* en phase gazeuse	polimerización *f* en fase gaseosa	polimerizzazione *f* in fase gas
gazeux (se)	gaseoso	gassoso
nitruré au gaz	nitrurado en gas	nitrurato con gas
point *m* d'injection	entrada *f;* punto *m* de inyección	punto *m* d'iniezione; orifizio *m* di entrata
pompe *f* à engrenage	bomba *f* de engranajes	pompa *f* a ruota dendata
crémaillère *f*	piñón *m* de cremallera	cremagliera *f*
engrenage *m*	rueda *f* dentada	ruota *f* dentata
gel *m*	gel *m*	gel *m*

Nr.	English	Deutsch
638	**gel coat**	Gelcoatschicht *f*
639	**gel content**	Gelanteil *m*
640	**gel permeation chromatography** *(GPC)*	Gel-Permeations-Chromatographie *f*
641	**gel point**	Gelpunkt *m*
642	**gelling**	Gelieren *n*
643	**gelling time**	Gelierzeit *f*
644	**general-purpose ...**	Allzweck ... *m*
645	**general-purpose rubber**	Allzweck-Kautschuk *m*
646	**glass beads**	Glaskugeln *fpl*
647	**glass content**	Glasgehalt *m*
648	**glass fibre**	Glasfaser *f*
649	**glass fibre mat**	Glasfasermatte *f*
650	**glass fibre reinforced plastic** *(GRP)*	Glasfaserkunststoff *m*; glasfaserverstärkter Kunststoff *m* (GFK)
651	**glass transition temperature**	Glastemperatur *f*

Français	*Español*	*Italiano*
gel coat *m*	gel coat *m*	resina *f* gel di finitura
pourcentage *f* de produit gélifié	contenido *m* en gel	contenuto *m* di gel
chromatographie *f* par perméation de gel	cromatografía *f* por permeabilidad de gel	cromatografia *f* a permeazione di gel
point *m* de gel	punto *m* de gelificación	punto *m* di gelificazione
gélification *f*	gelificación *f*	gelificazione *f*
temps *m* de gel	tiempo *m* de gelificación	tempo *m* di gelificazione
standard *m* …	uso *m* universal …	usi *mpl* generali …
caoutchouc *m* standard	caucho *m* de usos generales	gomma *f* standard
billes *fpl* de verre	bolas *fpl* de vidrio	sfere *fpl* di vetro
contenu *m* en verre	contenido *m* de fibra de vidrio	contenuto *m* di vetro
fibre *f* de verre	fibra *f* de vidrio	fibra *f* di vetro
mat *m* de fibres de verre	mat *m* de vidrio	stuoia *f* di vetro; mat *m* di vetro
plastique *m* renforcé fibres de verre	plástico *m* reforzado con fibra de vidrio	materie *fpl* plastiche rinforzate con fibre di vetro
température *f* de transition vitreuse	temperatura *f* de transición vítrea	temperatura *f* di transizione vetrosa

Nr.	English	Deutsch
652	**glassy state**	Glaszustand *m*
653	**gloss**	Glanz *m*
654	**glycerol**	Glycerin *n*
655	**graft polymer/copolymer**	Pfropf-Polymer/ Copolymer *n*
656	**graft polymerisation**	Pfropf(co)polymerisation *f*
657	**grafting**	Pfropfung *f*
658	**grain**	Körnung *f*
659	**granulator**	Schneidmühle *f*
660	**graphite**	Graphit *m*
661	**gravimetric feeding**	gravimetrisches Dosieren *n*
662	**gravure printing**	Tiefdruck *m*
663	**grey streaks**	Grauschlieren *fpl*
664	**grinder**	Mühle *f*
665	**grinding**	Mahlen *n*
666	**groove**	Nut *f*
667	**grooved**	genutet

Français	*Español*	*Italiano*
état *m* vitreux	estado *m* vítreo	stato *m* di vetrificazione
brillant *m*	brillo *m*	brillantezza *f*
glycérine *f*	glicerina *f*	glicerina *f*
polymère/ copolymère *m* greffé	polímero/copolímero *m* de injerto	polimero/copolimero *m* aggraffato
polymérisation *f* par greffage	polimerización *f* de injerto	polimerizzazione *f* per aggraffaggio
greffage *m*	injerto *m*	aggraffaggio *m*
grain *m*	gránulo *m*	grana *f*
broyeur *m*	granulador *m*	granulatore *m*
graphite *m*	grafito *m*	grafite *f*
dosage *m* gravimétrique	dosificación *f* gravimétrica	alimentazione *f* gravimetrica
héliogravure *f*	impresión *f* por huecograbado	stampa *f* a intaglio
traces *fpl* grises	vetas *fpl* grises	segni *mpl* grigi
broyeur *m*	molino *m*	mulino *m*
broyage *m*	moledura *f*; trituración *f*	macinazione *f*
rainure *f*	ranura *f*	scanalatura *f*
rainuré	ranurado	scanalato

Nr.	English	Deutsch
668	**grooved barrel extruder**	Nutenextruder *m*
669	**grooved bushing**	Nutbuchse *f*
670	**group transfer polymerisation** *(GTP)*	Gruppen-Transfer-Polymerisation *f*
671	**guide pin**	Führungsstift *m;* Führungssäule *f*
672	**guide value**	Richtwert *m*
673	**gutta-percha**	Guttapercha *n*
674	**hammer mill**	Hammermühle *f*
675	**hand lay-up**	Handlaminieren *n*
676	**hand mould**	Handform *f*
677	**handling**	Handhabung *f*
678	**handling device**	Handhabungsgerät *n*
679	**hard chrome plating**	Hartverchromen *n*
680	**hard-faced**	gepanzert
681	**hard rubber**	Hartgummi *m*

Français	*Español*	*Italiano*
extrudeuse *f* à cylindre rainuré	extrusora *f* con encamisado ranurado	estrusore *m* a cilindro scanalato
douille *f* rainurée	casquillo *m* ranurado; camisa *f* ranurada	cilindro *m* scanalato
polymérisation *f* par transfert de groupe	polimerización *f* por transferencia de grupos	polimerizzazione *f* a trasferimento di gruppo
broche *f* de guidage	espiga *f* de guía; perno *m* de guía	perno *m* di centratura; colonna *f* di guida
valeur *f* de consigne	valor *m* aproximado de orientación	valore *m* indicativo
gutta-percha *m*	gutapercha *f*	guttaperca *f*
broyeur *m* à couteaux	molino *m* de mazos	mulino *m* a lame
drapage *f* manuel	laminado *m* a mano	spalmatura *f* a mano
moulage *m* manuel	molde *m* manual	stampo *m* manuale
manipulation *f*	manejo *m*	maneggiamento *m*
manipulateur *m*	instrumento *m* de operación	apparecchiatura *f* di manipolazione
chromage *m*	cromado *m* duro	cromatura *f* a spessore
blindé	templado	con riporto duro
caoutchouc *m* rigide	goma *f* dura; ebonita *f*	gomma *f* indurata; ebanite *f*

Nr.	English	Deutsch
682	hardener	Härter *m*
683	hardness	Härte *f*
684	hardness test	Härteprüfung *f*
685	haul-off *(extr.)*	Abzug *m*
686	haul-off ratio	Abzugsverhältnis *n*
687	haze	Trübung *f*
688	head-tail polymerisation	Kopf-Schwanz-Polymeri-sation *f*
689	heat ageing	Wärmealterung *f*
690	heat distortion temperature *(HDT)*	Formbeständigkeit *f* in der Wärme; Wärmeformbeständigkeit *f*
691	heat exchanger	Wärmetauscher *m*
692	heat resistance	Wärmeformbeständigkeit *f*
693	heat resistant	wärmeformbeständig
694	heat-sealing	Heißsiegeln *n*
695	heat stabiliser	Wärmestabilisator *m*

Français	*Español*	*Italiano*
durcisseur *m*	endurecedor *m*	induritore *m*
dureté *f*	dureza *f*	durezza *f*
essai *m* de dureté	ensayo *m* de dureza	prova *f* di durezza
tirage *m*	equipo *m* de tracción	traino *m* (nell'estrusione)
taux *m* de tirage	relación *f* de estiraje	percentuale *f* di traino
trouble *m*	opacidad *f*	opacità *f*
polymérisation *f* tête-bêche	polimerización *f* cabeza a cola	polimerizzazione *f* testa-coda
vieillissement *m* à la chaleur	envejecimiento *m* por calor	invecchiamento *m* a caldo
température *f* de déformation à la chaleur	temperatura *f* de distorsión por el calor	temperatura *f* di deformazione al calore
échangeur *f* de chaleur	intercambiador *m* de calor	scambiatore *m* di calore
résistance *f* à la chaleur	resistencia *f* al calor	resistenza *f* termica
résistant à la chaleur; thermostable	resistente al calor	termoresistente
soudage *m* à la chaleur	soldadura *f* por calor; sellado *m* en caliente	termosaldatura *f*
stabilisant *f* chaleur	estabilizante *m* térmico	stabilizzatore *m* di calore

Nr.	English	Deutsch
696	**heat stability**	Wärmestabilität *f*
697	**heat transfer**	Wärmeübergang *m*
698	**heat transfer coefficient**	Wärmedurchgangszahl *f*
699	**heat treatment**	Wärmebehandlung *f*
700	**heat, to**	beheizen
701	**heated tool welding**	Heizelementschweißen *n*
702	**heater band**	Heizband *n*
703	**heating capacity**	Heizleistung *f*
704	**heating channel**	Heizkanal *m*
705	**heating equipment**	Temperiergerät *n*
706	**heating rolls**	Heizwalzen *fpl*
707	**heating time**	Aufheizzeit *f*
708	**heating zone**	Temperierzone *f;* Heizzone *f*

Français	*Español*	*Italiano*
stabilité *f* à la chaleur	resistencia *f* al calor	resistenza *f* termica
transfert *m* de chaleur	transmisión *f* de calor	cessione *m* di calore
coefficient *m* de transfert de chaleur	coeficiente *m* de conductividad térmica	coefficiente *m* trasferimento di calore
traitement *m* thermique	tratamiento *m* térmico	trattamento *m* termico
chauffer	calentar	riscaldare
soudage *m* par éléments chauffants	soldadura *f* por elementos calientes	saldatura *f* con barra calda
collier *m* chauffant	banda *f* calefactora	banda *f* riscaldante
capacité *f* de chauffage	capacidad *f* calefactora	capacità *f* di riscaldamento
canal *m* chauffant	canal *m* de calentamiento	canale *m* di riscaldamento
équipement *m* de chauffage	equipo *m* calefactor	attrezzatura *f* di riscaldamento
rouleaux *mpl* chauffants	cilindros *mpl* calefactores	cilindri *mpl* riscaldanti
temps *m* de chauffage	tiempo *m* de calefacción	tempo *m* di riscaldamento
zone *f* de chauffe	zona *f* de calefacción	zona *f* di riscaldamento

Nr.	English	Deutsch
709	**high-density polyethylene** *(PE-HD)*	Niederdruck-Polyethylen *n;* Polyethylen *n* hoher Dichte
710	**high-frequency heating**	Hochfrequenzerwärmung *f*
711	**high-frequency welding**	Hochfrequenzschweißen *n*
712	**high-gloss**	Hochglanz *m*
713	**high-impact polystyrene** *(PS-HI)*	Polystyrol *n,* schlagzähes
714	**high-impact resistant**	schlagfest; schlagzäh
715	**high-performance plastics**	Hochleistungs-kunststoffe *mpl*
716	**high-speed extruder**	Schnellläufer-Extruder *m*
717	**holding pressure**	Nachdruck *m*
718	**holding pressure stroke**	Nachdruckweg *m*
719	**hollow article**	Hohlkörper *m*
720	**homogenisation**	Homogenisierung *f*
721	**homopolymer**	Homopolymer *n*

Français	Español	Italiano
polyéthylène *m* haute densité (PEHD)	polietileno *m* de alta densidad	polietilene *m* alta densità (PE a.d.)
chauffage *m* haute fréquence	calentamiento *m* por alta frecuencia	riscaldamento *m* ad alta frequenza
soudage *m* à haute fréquence	soldadura *f* por alta frecuencia	saldatura *f* ad alta frequenza
grande brillance (high gloss) *f*	alto brillo *m*	brillantezza *f* elevata
polystyrène *m* choc	poliestireno *m* alto impacto	polistirene *m* resistente all'urto
résistant au choc	resistente al impacto	resistente all'urto
plastiques *mpl* hautes performances	plásticos *mpl* de altas prestaciones	materie plastiche *fpl* a prestazioni elevate
extrudeuse *f* à haute vitesse	extrusora *f* de alta velocidad	estrusore *m* ad alta velocità
pression *f* de maintien	presión *f* posterior; presión de mantenimiento	pressione *f* di mantenimento
course *f* de pression de maintien	recorrido *m* de la presión de mantenimiento	corsa *f* della pressione di mantenimento
corps *m* creux	artículo *m* hueco	corpo *m* cavo
homogénéisation *f*	homogenización *f*	omogeneizzazione *f*
homopolymère *m*	homopolímero *m*	omopolimero *m*

Nr.	English	Deutsch
722	**honeycomb material**	Honigwaben-Baustoff *m*
723	**hoop stress**	Umfangsspannung *f*
724	**hopper**	Trichter *m*
725	**hopper dryer**	Trichtertrockner *m*
726	**hopper heater**	Trichterheizung *f*
727	**hopper loader**	Einfüll-Leitung *f*
728	**hose**	Schlauch *m*
729	**hot curing**	Wärmehärtung *f*
730	**hot gas welding**	Heißgasschweißen *n;* Warmgasschweißen *n*
731	**hot melt adhesive**	Schmelzkleber *m*
732	**hot plate**	Heizelement *n*
733	**hot plate butt welding**	Heizelementstumpfschweißen *n;* Heizspiegelschweißen *n*
734	**hot press moulding**	Warmpressverfahren *n*
735	**hot runner**	Heißkanal *m*

Français	*Español*	*Italiano*
matériau *m* en nid d'abeille	material *m* de panel de abeja	materiale *m* a nido d'ape
contrainte *f* circulaire	tensión *f* circular	tensione *f* circolare
trémie *f*	tolva *f* de alimentación	tramoggia *f*
sécheur *m* de trémie *f*	secador *m* de tolva	essiccatore *m* da tramoggia
chauffeur *m* de trémie	calefactor *m* de la tolva	riscaldatore *m* da tramoggia
chargeur *m* de trémie	alimentador *m* de tolva	alimentatore *m* da tramoggia
tuyau *m*	tubo *m* flexible	tubo *m* flessibile
réticulation *f* par la chaleur	curado *m* en caliente	vulcanizzazione *f* a caldo
soudage *m* aux gaz chauds	soldadura *f* por gas caliente	saldatura *f* a gas caldo
adhésif *m;* hot melt adhésif *m* thermofusible	adhesivo *m* de fusión	pellicola *f* adesiva a caldo
élément *m* chauffant	placa *f* de calefacción	piastra *f* riscaldante
soudage *m* par éléments chauffants	soldadura *f* por elementos calientes	saldatura *f* di testa a attrezzo caldo
moulage *m* à chaud à la presse	prensado *m* en caliente	stampaggio *m* a barra calda
canal *m* chauffant	canal *m* caliente	canale *m* caldo

Nr.	English	Deutsch
736	**hot runner mould**	Heißkanalwerkzeug *n*
737	**hot runner nozzle**	Heißkanaldüse *f*
738	**hot stamping**	Heißprägen *n*
739	**hot wire cutter**	Glühdrahtabschneider *m*
740	**hot wire welding**	Heizdrahtschweißen *n;* Glühdrahtschweißen *n*
741	**hybrid**	hybrid
742	**hybrid drive**	Hybridantrieb *m*
743	**hydraulic clamp**	hydraulische Schließeinheit *f*
744	**hydraulic ejector**	hydraulischer Auswerfer *m*
745	**hydraulic pump**	Hydraulikpumpe *f*
746	**hydraulic system**	Hydrauliksystem *n*
747	**hydrogenation**	Hydrierung *f*
748	**hydrolysis**	Hydrolyse *f*
749	**hydrophilic**	hydrophil
750	**hydrophobic**	hydrophob

Français	Español	Italiano
moule *m* à canaux chauds	molde *m* de canal caliente	stampo *m* a canali caldi
buse *f* pour canal chaud	boquilla *f* del canal caliente	ugello *m* a canale caldo
impression *f* à chaud	estampación *f* en caliente	marcatura *f* a caldo
coupeur *m* au fil chaud	cortador *m* de hilo caliente	taglierina *f* a filo caldo
soudage *m* au cordon chauffant	soldadura *f* por alambre caliente	saldatura *f* a filo caldo
hybride	híbrido	ibrido
entraînement *m* hybride	accionamiento *m* híbrido	comando *m* ibrido
fermeture *f* hydraulique	cierre *m* hidraulíco	chiusura *f* idraulica
éjecteur *m* hydraulique	expulsor *m* hidraulico	espulsore *m* idraulico
pompe *f* hydraulique	bomba *f* hidráulica	pompa *f* idraulica
système *m* hydraulique	sistema *m* hidráulico	sistema *m* idraulico
hydrogénation *f*	hidrogenación *f*	idrogenazione *f*
hydrolyse *f*	hidrólisis *f*	idrolisi *f*
hydrophile	hidrófilo	idrofilo
hydrophobe	hidrófobo	idrorepellente

Nr.	English	Deutsch
751	hygroscopic	hygroskopisch
752	hysteresis	Hysterese *f*
753	idling roller	Tänzerwalze *f*
754	ignition temperature	Zündpunkt *m*
755	imidisation	Imidisierung *f*
756	impact modifier	Schlagzähigkeits-verbesserer *m*
757	impact strength	Schlagzähigkeit *f*
758	impact test	Schlagversuch *m*
759	impregnate, to	imprägnieren
760	impregnating bath	Tränkbad *n*
761	impregnating resin	Tränkharz *n*
762	impulse welding	Impulsschweißen *n*
763	impulse welding (thermal)	Wärmeimpulsschweißen *n*
764	in-mould coating *(IMC)*	In-Mould-Lackieren *n*

Français	*Español*	*Italiano*
hygroscopique	higroscópico	igroscopico *m*
hystérésis *f*	histéresis *f*	isteresi *f*
cylindre *m* oscillant	rodillo *m* palpador	rullo *m* di guida
température *f* d'incandescence	temperatura *f* de ignición	temperatura *f* di ignizione
imidisation *f*	imidización *f*	immidizzazione *f*
modifiant *m* au choc	modificador *m* de impacto	agente *m* antiurto
résistance *f* au choc	resistencia *f* al impacto	resistenza *f* all'urto
essai *m* au choc	ensayo *m* al impacto	prova *f* d'urto
imprégner	impregnar	impregnare
bain *m* d'imprégnation	baño *m* de impregnación	bagno *m* di impregnazione
résine *f* d'imprégnation	resina *f* de impregnación	resina *f* impregnante
soudage *m* par impulsion	soldadura *f* por impulsión	saldatura *f* a impulsi
soudage *m* par impulsions thermiques	soldadura *f* por impulso térmico	termosaldatura *f* a impulsi
revêtement *m* en moule	técnica *f* de pintado en el molde (IMC)	rivestimento *m* nello stampo

Nr.	English	Deutsch
765	**in-mould decoration** *(IMD)*	In-Mould-Dekorieren *n*
766	**in-mould labelling** *(IML)*	Etikettenhinterspritzen *n;* IML-Technik *f*
767	**in-mould surface decoration**	Dekorhinterspritzen *n*
768	**in-situ foam**	Ortsschaum *m*
769	**incompatibility**	Unverträglichkeit *f*
770	**indentation resistance**	Eindruckwiderstand *m;* Eindruckfestigkeit *f*
771	**induction period**	Induktionsperiode *f*
772	**infusible; non-fusible**	unschmelzbar
773	**inherent rigidity**	Formsteifigkeit *f*
774	**inhibitor**	Inhibitor *m*
775	**injection blow moulding**	Spritzblasen *n*
776	**injection capacity**	Hubvolumen *n;* Schussvolumen *n*
777	**injection compression moulding**	Spritzprägen *n*

Français	*Español*	*Italiano*
décoration *f* en moule	técnica *f* de decorado en el molde (IMD)	decorazione *f* nello stampo
étiquetage *m* en moule	etiquetado *m* en el molde (IML)	etichettatura *f* nello stampo
décoration *f* surfacique en moule	decorado *m* en la superficie del molde	decorazione *f* superficiale nello stampo
mousse *f* in situ	espumación *f* «in situ»	espansione *f* in loco
incompatibilité *f*	incompatibilidad *f*	incompatibilità *f*
résistance *f* à l'indentation	resistencia *f* a la penetración	resistenza *f* alla penetrazione
période *f* d'induction	período *m* de inducción	periodo *m* di induzione
infusible	infusible	infusibile
rigidité *f* intrinsèque	rigidez *f* inherente	rigidità *f* intrinseca
inhibiteur *m*	inhibidor *m*	inibitore *m*
injection *f* soufflage	moldeo *m* por inyección-soplado	stampaggio *m* per soffiaggio a iniezione
capacité *f* d'injection	capacidad *f* de inyección	capacità *f* di iniezione
moulage *m* par injection-compression	inyección *f* por compresión	stampaggio *m* per compressione iniezione

Nr.	English	Deutsch
778	injection mould	Spritzgießwerkzeug *n*
779	injection moulded part	Spritzgussteil *n*
780	injection moulding	Spritzgießen *n*
781	injection moulding faults	Spritzgießfehler *mpl*
782	injection moulding machine	Spritzgießmaschine *f*
783	injection nozzle	Einspritzdüse *f*
784	injection plunger	Spritzkolben *m*
785	injection power	Einspritzleistung *f*
786	injection pressure	Einspritzdruck *m*
787	injection process	Injektionsverfahren *n*
788	injection rate	Einspritzgeschwindigkeit *f*; Einspritzstrom *m*
789	injection stretch blow moulding	Spritz-Streckblasen *n*

Français	Español	Italiano
moule *m* d'injection	molde *m* de inyección	stampo *m* per iniezione
pièce *f* moulée par injection	pieza *f* moldeada por inyección	pezzo *m* stampato a iniezione
moulage *m* par injection	moldeo *m* por inyección	stampaggio *m* a iniezione
défauts *mpl* de moulage par injection	defectos *mpl* de la inyección	difetti *mpl* nello stampaggio a iniezione
machine *f* à injecter	máquina *f* de moldeo por inyección; inyectora *f*	pressa *f* per iniezione
buse *f* d'injection	boquilla *f* de inyección	ugello *m* di iniezione
piston *m* d'injection	émbolo *m* de inyección	pistone *m* di iniezione
puissance *f* d'injection	potencia *f* de inyección	potenza *f* di iniezione
pression *f* d'injection	presión *f* de inyección	pressione *f* di iniezione
procédé *m* d'injection	proceso *m* de inyección	processo *m* di iniezione
vitesse *f* d'injection	velocidad *f* de inyección	tasso *m* di iniezione
injection *f* soufflage par biorientation	moldeo *m* por inyección-estirado por soplado	iniezione *f* soffiaggio con biorientazione

Nr.	English	Deutsch
790	**injection stroke**	Einspritzweg *m*
791	**injection time**	Einspritzzeit *f*
792	**injection unit**	Spritzeinheit *f*
793	**ink-jet printing**	Tintenstrahl-Beschriften *n*
794	**inlay printing**	Farbprägen *n*
795	**inorganic**	anorganisch
796	**insert**	Einpressteil *n;* Einlegeteil *n*
797	**insert moulding**	Insert-Technik *f;* Umspritzen *n*
798	**insert-placing robot**	Einlegeroboter *m*
799	**insulating bushing**	Isolierbuchse *f*
800	**insulating material**	Dämmstoff *m*
801	**insulating varnish**	Isolierlack *m*
802	**insulation**	Isolierung *f*
803	**integral hinge**	Filmscharnier *n*

Français	Español	Italiano
course *f* d'injection	recorrido *m* del flujo de inyección	corsa *f* di iniezione
temps *m* d'injection	tiempo *m* de inyección	tempo *m* di iniezione
unité *f* d'injection	unidad *f* de inyección	unità *f* di iniezione
impression *f* par jet d'encre	impresión *f* por chorro de tinta	stampa *f* a getto di inchiostro
impression *f* en couleur	estampado *m* en colores	stampa *f* a intarsio
inorganique (minéral)	inorgánico	inorganico
insert *m*	inserto *m*	inserto *m*
moulage *m* d'insert	moldeo *m* con inserción	stampaggio *m* con inserto
robot *m* poseur d'inserts	robot *m* para la colocación de insertos	manipolatore *m* per posizionamento inserti
bague *f* d'isolation	manguito *m* aislante	boccola *f* isolante
matériau *m* isolant	material *m* aislante	materiale *m* isolante
vernis *m* isolant	barniz *m* aislante	vernice *f* isolante
isolation *f*	aislamiento *m*	isolamento *m*
lisière *f* intégrale	bisagra *f* integral de lámina	cerniera *f* integrale

Nr.	English	Deutsch
804	**interface**	Schnittstelle *f*
805	**interfacial polymerisation**	Grenzflächen-Polymerisation *f*
806	**interlaminar**	interlaminar
807	**interlock**	Verriegelung *f*
808	**internal calibration**	Innenkalibrierung *f*
809	**internal cooling**	Innenkühlung *f*
810	**internal mixer**	Innenmischer *m*
811	**internal pressure**	Innendruck *m*
812	**internal stress**	Eigenspannung *f*
813	**internal thread**	Innengewinde *n*
814	**interpenetrating network**	durchdringendes Netzwerk *n*
815	**intrinsic viscosity**	Grenzviskosität *f*
816	**ionic polymerisation**	ionische Polymerisation *f*
817	**ionitrided**	ionitriert
818	**ionomer**	Ionomer *n*
819	**irradiation**	Bestrahlung *f*

Français	*Español*	*Italiano*
interface *f*	interface *m*	interfaccia *f*
polymérisation *f* interfaciale	polimerización *f* de interfases	polimerizzazione *f* interfacciale
interlaminaire	interlaminar	interlaminare
enclavement *m*	enclavamiento *m*	interblocco *m*
calibrage *m* interne	calibración *f* interna	calibrazione *f* interna
refroidissement *m* interne	enfriamiento *m* interno	raffreddamento *m* interno
mélangeur *m* interne	mezclador *m* interno	mescolatore *m* interno
pression *f* interne	presión *f* interna	pressione *f* interna
contrainte *f* interne	tensión *f* interna	tensione *f* interna
filet *m* intérieur	rosca *f* interna	filettatura *f* interna
réseau *m* interpénétré	retículo *m* penetrante	reticolo *m* interpenetrante
viscosité *f* intrinsèque	número *m* de viscosidad límite	viscositá *f* intrinseca
polymérisation *f* ionique	polimerización *f* iónica	polimerizzazione *f* ionica
nitruré par ions	ionitrurado	nitrurato con ioni
ionomère *m*	ionómero *m*	ionomero *m*
irradiation *f*	irradiación *f*	irradiazione *f*

Nr.	English	Deutsch
820	**isochronous stress-strain curve**	isochrone Spannungs-Dehnungskurve *f*
821	**isocyanate**	Isocyanat *n*
822	**isocyanurate**	Isocyanurat *n*
823	**isomorphism**	Isomorphie *f*
824	**isotactic**	isotaktisch
825	**isotropic**	isotrop
826	**isotropy**	Isotropie *f*
827	**jacket**	Mantel *m*
828	**jaw chuck**	Backenfutter *n*
829	**jetting**	Freistrahlbildung *f*
830	**joining**	Fügen *n*
831	**joining pressure**	Fügedruck *m*
832	**joint**	Verbindung *f*
833	**k-value**	k-Wert *m*
834	**kinetics**	Kinetik *f*
835	**kneader**	Kneter *m*

Français	*Español*	*Italiano*
courbe *f* de contrainte déformation isochrone	curva *f* isocrona esfuerzo-alargamiento	curva *f* isocrona di sollecitazione-deformazione
isocyanate *m*	isocianato *m*	isocianato *m*
isocyanurate *m*	isocianurato *m*	isocianurato *m*
isomorphisme *m*	isomorfía *f*	isomorfismo *m*
isotactique	isotáctico	isotattico
isotropique	isótropo	isotropo
isotropie *f*	isotropía *f*	isotropia *f*
enveloppe *f*	camisa *f*	camicia *f*
mandrin *m* à mâchoire	mandril *m* de mordazas	mandrino *m* a ganasce
jet *m* libre	jetting *m*	getto *m* libero
assemblage *m*	ensamblaje *m*	giunzione *f*
pression *f* d'assemblage	presión *m* de ensamblaje	pressione *f* di giunzione
joint *m*	unión *f*	giunto *m*
k-wert *m*	valor-k *m*	fattore-k *m*
cinétique *f*	cinética *f*	cinetica *f*
malaxeur *m*	amasadora *f*; amasador *m*	impastatrice *f*

Nr.	English	Deutsch
836	**knit line**	Schweißnaht *f*
837	**knitted fabric**	Gewirke *n*
838	**labelling**	Kennzeichnung *f*
839	**lacquer**	Lack *m*
840	**ladder polymers**	Leiterpolymere *npl*
841	**laminar flow**	Quellfluss *m;* laminares Fließen *n*
842	**laminate**	Schichtstoff *m;* Laminat *n*
843	**laminated paper**	Hartpapier *n*
844	**laminating**	Kaschieren *n*
845	**laminating roll**	Kaschierwalzwerk *n*
846	**land** *(of a screw)*	Steg *m;* Schneckensteg *m*
847	**lap-welding**	Überlappschweißen *n*
848	**laser cutting**	Laserschneiden *n*
849	**laser marking**	Laser-Beschriften *n*
850	**laser welding**	Laserschweißen *n*

Français	*Español*	*Italiano*
ligne *f* de soudure	línea *f* de soldadura	linea *f* di filatura
tissu *m* tricoté	género *m* de punto	tessuto *m* a maglia
étiquetage *m*	etiquetado *m*	etichettatura *f*
peinture *f*	laca *f;* esmalte *m*	lacca *f;* vernice *f*
polymères *mpl* en échelle	polímeros *mpl* de escalera	polimero *m* a scala
flux *m* laminaire	flujo *m* laminar	flusso *m* laminare
stratifié *m;* laminé *m*	laminado *m;* estratificado *m*	stratificato *m;* accoppiato *m;* laminato *m*
papier *m* couché	estratificado *m* de papel	carta *f* accoppiata
couchage *m;* laminage *m*	laminado *m;* doblado *m*	stratificazione *f*
cylindre *m* de couchage	cilindro *m* de laminación	cilindro *m* d'accoppiamento, di laminazione
largeur *f* du filet	anchura *f* del filete	volume *m* del canale
soudage *m* par recouvrement	soldadura *f* a solape	saldatura *f* a sovrapposizione
découpe *m* au laser	corte *m* por láser	taglio *m* laser
marquage *m* au laser	impresión *f* por láser	marcatura *f* a laser
soudage *m* laser	soldadura *f* por láser	saldatura *f* a laser

Nr.	English	Deutsch
851	**lateral adjustment**	Seitenverstellung *f*
852	**latex**	Latex *m*
853	**lead stabiliser**	Bleistabilisator *m*
854	**leakage flow**	Leckströmung *f*
855	**leakage test**	Dichtheitsprüfung *f*
856	**lidding film**	Deckelfolie *f*
857	**light resistance**	Lichtbeständigkeit *f*
858	**light scattering**	Lichtstreuung *f*
859	**light stabiliser**	Lichtschutzmittel *n*
860	**light transmission; transparency**	Lichtdurchlässigkeit *f*
861	**limiting oxygen index** *(LOI)*	Sauerstoffindex *m*
862	**lining**	Auskleidung *f*
863	**liquid**	flüssig
864	**liquid crystal polymers** *(LCP)*	flüssigkristalline Polymere *npl*

Français	Español	Italiano
réglage latéral *m*	ajuste *m* lateral	regolazione *f* laterale
latex *m*	látex *m*	lattice *m*
stabilisant *m* au plomb	estabilizante *m* de plomo	stabilizzante *m* al piombo
fuite *f*	flujo *m* de pérdidas; flujo *m* de escape	flusso *m* di rigurgito
test *m* d'étanchéité	ensayo *m* de estanqueidad	prova *f* di tenuta
film *m* housse	hoja *m* de cierre	film *m* sigillante
résistance *f* à la lumière	resistencia *f* a la luz	resistenza *f* alla luce
diffusion *f* de la lumière	dispersión *f* de la luz	dispersione *f* di luce
stabilisant *m* à la lumière	estabilizante *m* a la luz	stabilizzante *m* alla luce
transmission *f* de la lumière; transparence	permeabilidad *f* a la luz	trasparenza *f* alla luce
indice *m* limite d'oxygène	índice *m* de oxígeno límite	indice *m* limite di ossigeno
revêtement *m*	revestimiento *m*	rivestimento *m*
liquide	líquido	liquido
polymères *mpl* à cristaux liquides	polímeros *mpl* de cristal líquido	polimeri *mpl* cristallini liquidi

Nr.	English	Deutsch
865	**liquid injection moulding** *(LIM)*	Spritzgießen *n* flüssiger Kunststoffe
866	**liquid resin press moulding**	Nasspressverfahren *n*
867	**liquid rubber**	Flüssigkautschuk *m*
868	**living polymers**	lebende Polymere *npl*
869	**loading chamber**	Füllraum *m*
870	**locating ring**	Zentrierring *m*
871	**locking force**	Zuhaltekraft *f*
872	**locking mechanism**	Schließvorrichtung *f;* Arretiervorrichtung *f*
873	**long-term flexural strength**	Dauerbiegefestigkeit *f*
874	**long-term stress**	Dauerbeanspruchung *f*
875	**long-term test**	Langzeitversuch *m*
876	**lost core**	Schmelzkern *m*
877	**low-density polyethylene** *(PE-LD)*	Hochdruck-Polyethylen *n;* Polyethylen *n* niedriger Dichte
878	**low-pressure moulding**	Niederdruck-Pressverfahren *n*

Français	*Español*	*Italiano*
moulage *m* par injection de liquide	moldeo *m* por inyección de líquido	stampaggio *m* a iniezione di liquido
moulage *m* de résine liquide à la presse	moldeo *m* por prensado de resina líquida	stampaggio *m* a pressione di resina liquida
caoutchouc *m* liquide	caucho *m* líquido	gomma *f* liquida
polymères *mpl* vivants	polímero *m* vivo	polimeri *mpl* vivi
chambre *f* de chargement	cámara *f* de transferencia	camera *f* di caricamento
anneau *m* de centrage	anillo *m* de centraje	anello *m* di centratura
force *f* de fermeture	fuerza *f* de cierre	forza *f* di chiusura
mécanisme *m* de fermeture	mecanismo *m* de cierre	meccanismo *m* di chiusura
résistance *f* à la flexion en continu	resistencia *f* a la fatiga por flexión	resistenza *f* alla sollecitazione ripetute
contrainte *f* en continu	esfuerzo *m* de fatiga	sollecitazione *f* ripetute
essai *m* à long terme	ensayo *m* de larga duración	prova *f* a lungo termine
noyau *m* fusible	núcleo *m* fundente	nucleo *m* a perdere
polyéthylène *m* basse densité (PEBD)	polietileno *m* de baja densidad	polietilene *m* bassa densità (PE b.d.)
moulage *m* basse pression	prensado *m* a baja presión	stampaggio *m* a bassa pressione

Nr.	English	Deutsch
879	**low-pressure process**	Niederdruckverfahren *n*
880	**low-profile resins** *(LP)*	Low-Profile-Harz *n*
881	**lubricant**	Gleitmittel *n*
882	**luminance curve**	Remissionskurve *f*
883	**lyotropic**	lyotrop
884	**machine, to**	bearbeiten
885	**machine base**	Maschinengestell *n*
886	**machine setter**	Einrichter *m*
887	**machine settings**	Maschineneinstellung *f*
888	**macromolecule**	Makromolekül *n*
889	**maintenance**	Wartung *f*
890	**mandrel**	Pinole *f;* Dorn *m*
891	**mandrel support**	Dornhalter *m*
892	**manifold**	Verteiler *m;* Schmelzeverteiler *m*
893	**marbling**	Marmorieren *n*

125 *marbling*

Français	Español	Italiano
procédé *m* basse pression	proceso *m* de baja presión	processo *m* a bassa pressione
résine *f* low profile	resina *f* low-profile (de baja contracción)	resina *f* a basso profilo
lubrifiant *m*	lubricante *m*	lubrificante *m*
courbe *f* de luminance	curva *f* de remisión	curva *f* di luminescenza
lyotrope (cristal liquide)	liotrópico	liotropico
usiner	mecanizar	lavorare all'utensile
châssis *m* de la machine	chasis *m* de la máquina	basamento *m* macchina
régleur *m*	operario *m* de la máquina	programmatore *m* macchina
réglage *m* de la machine	parámetros *m* de la máquina	messa a punto *f* della macchina
macromolécule *f*	macromolécula *f*	macromolecola *f*
entretien *m*	mantenimiento *m*	manutenzione *f*
mandrin *m;* douille *f*	mandril *m*	mandrino *m*
porte-poinçon *m*	portamandril *m*	anello *m* a razze
répartiteur *m*	distribuidor *m*	collettore *m*
marbrure *f*	efecto *m* mármol	marmorizzazione *f*

Nr.	English	Deutsch
894	**Martens temperature**	Martenszahl *f*
895	**mass polymerisation**	Massepolymerisation *f*
896	**masterbatch**	Masterbatch *n*
897	**mastication**	Mastikation *f*
898	**mat**	Matte *f*
899	**mat moulding**	Mattenpressverfahren *n*
900	**mat reinforcement**	Mattenverstärkung *f*
901	**material change-over**	Materialwechsel *m*
902	**material degradation**	Materialzersetzung *f*
903	**materials testing**	Werkstoffprüfung *f;* Materialprüfung *f*
904	**matrix**	Matrix *f*
905	**maturing time**	Reifungszeit *f*
906	**maximum allowable concentration** *(MAC)*	maximale Arbeitsplatzkon-zentration *f* (MAK-Wert *m*)
907	**mechanical blowing**	Begasen *n*

Français	*Español*	*Italiano*
température *f* Martens	temperatura *f* Martens	temperatura *f* Martens
polymérisation *f* en masse	polimerización *f* en masa	polimerizzazione *f* in massa
mélange-maître *m*	mezcla *f* básica; mezcla *f* madre	miscela *f* madre
mastication *f*	masticación *f*	masticazione *f*
mat *m*	fieltro *m*	mat *m*
moulage *m* au contact par mat	moldeo *m* de fieltros por prensado	stampaggio *m* mat; confezionamento *m* a umido
mat *m* de renfort	fietro *m* de refuerzo	mat *m* di rinforzo
changement *m* de matière	cambio *m* de material	cambio *m* del materiale
dégradation *f* de la matière	degradación *f* del material	degradazione *f* del materiale
essai *m* de matériaux	ensayo *m* de materiales	prova *f* materiali
matrice *f*	matriz *f*	matrice *f*
temps *m* de mûrissement	tiempo *m* de maduración	tempo *m* di maturazione
concentration *f* maximale sur le lieu de travail	concentración *f* máxima admisible (MCA)	concentrazione *f* massima ammissibile (CMA)
insufflation *f* de gaz	soplado *m* mecánico	soffiaggio *m* meccanico

Nr.	English	Deutsch
908	**mechanical properties**	mechanische Eigenschaften *fpl*
909	**melamine resin**	Melaminharz *n*
910	**melt**	Schmelze *f*
911	**melt accumulator**	Schmelzespeicher *m*
912	**melt cushion**	Schmelzepolster *n;* Massepolster *n*
913	**melt flow index** *(MFI)*	Schmelzindex *m*
914	**melt flow-way system**	Schmelzeleitsystem *n*
915	**melt fracture**	Schmelzbruch *m*
916	**melt front**	Fließfront *f*
917	**melt homogeneity**	Schmelzehomogenität *f*
918	**melt pressure**	Schmelzedruck *m;* Massedruck *m*
919	**melt processing**	Schmelzeverarbeitung *f*

Français	*Español*	*Italiano*
propriétés *fpl* mécaniques	propiedades *mpl* mecánicas	proprietà *fpl* meccaniche
résine *f* mélamine	resina *f* de melamina	resina *f* melaminica
matière *f* fondue	masa *f* fundida; fundido *m*	materiale *f* fuso
accumulateur *m* de matière	acumulador *m* del fundido	accumulatore *m* della massa fusa
matelas *m* de matière	cojín *m* de masa fundida	cuscino *m* di massa fusa
indice *m* de fluidité	índice *m* de fluidez	indice *m* di fluidità a caldo
système *m* distributeur de la matière plastique fondue	conducto *m* de la masa fundida	sistema *m* di scorrimento del fuso
rupture *f* de l'écoulement	rotura *f* de flujo	rottura *f* della massa fusa
front *m* de matière fondue	frente *m* del fundido	fronte *m* del fuso
homogénéité *f* de la matière	homogeneidad *f* de la masa fundida	omogeneità *f* della massa
pression *f* de la matière fondue	presión *f* del fundido	pressione *f* di fusione
transformation *f* de la matière fondue	transformación *f* por fusión	trasformazione *f* per fusione

Nr.	English	Deutsch
920	**melt spinning**	Schmelzespinnen *n*
921	**melt strength**	Schmelzefestigkeit *f*
922	**melt temperature**	Schmelzetemperatur *f;* Massetemperatur *f*
923	**melt viscosity**	Schmelzeviskosität *f*
924	**melting core technology**	Schmelzkerntechnik *f*
925	**melting heat**	Schmelzwärme *f*
926	**melting point**	Schmelzpunkt *m*
927	**melting range**	Schmelzbereich *m*
928	**melting temperature**	Schmelztemperatur *f*
929	**mesotactic**	mesotaktisch
930	**metal deactivator**	Metalldesaktivator *m*
931	**metal detector**	Metalldetektor *m*
932	**metal injection moulding**	Metallspritzguss *m*
933	**metallising**	Metallisieren *n*

Français	*Español*	*Italiano*
malaxage *m* de la matière fondue	hilatura *f* de fusión	filatura *f* da fuso
tenue *f* de la matière à l'état fondu	resistencia *f* del fundido	resistenza *f* del fuso
température *f* de la matière fondue	temperatura *f* de la masa fundida	temperatura *f* della massa
viscosité *f* de la matière fondue	viscosidad *f* del fundido	viscosità *f* della massa
technique *f* à noyaux fusibles	técnica *f* de noyo fundible	tecnologia *f* del nucleo a perdere
chaleur *f* de fusion	calor *m* de fusión	calore *f* di fusione
point *m* de fusion	punto *m* de fusión	punto *m* di fusione
zone *f* de fusion	intervalo *m* de fusión	intervallo *m* di fusione
température *f* de fusion	temperatura *f* de fusión	temperatura *f* di fusione
mésotactique	mesotáctico	mesotattico
désactivateur *m* métallique	desactivador *m* metálico	disattivatore *m* di metalli
détecteur *m* de métaux	detector *m* de metales	rilevatore *m* di metalli
injection *f* de métal	moldeo *m* por inyección de metal	stampaggio *m* a iniezione di metalli
métallisation *f*	metalización *f*	metallizzazione *f*

Nr.	English	Deutsch
934	**metathesis**	Metathese *f*
935	**meter, to**	dosieren
936	**metering pump**	Dosierpumpe *f*
937	**metering section** *(screw)*	Ausstoßzone *f;* Meteringzone *f*
938	**metering stroke**	Dosierweg *m*
939	**metering unit**	Dosiervorrichtung *f*
940	**methacrylate**	Methacrylat *n*
941	**methyl methacrylate**	Methylmethacrylat *n*
942	**mica**	Glimmer *m*
943	**micell**	Mizelle *f*
944	**microbial resistance**	Beständigkeit *f* gegen Mikroorganismen
945	**microcellular rubber**	Porengummi *m*
946	**microscopy**	Mikroskopie *f*
947	**microsphere**	Mikrokugel *f*
948	**microstructure**	Mikrostruktur *f*

Français	*Español*	*Italiano*
métathèse *f*	metatesis *f*	metatesi *f*
doser	dosificar	dosare; alimentare
pompe *f* doseuse	bomba *f* de dosificación	pompa *f* di alimentazione; pompa *f* dosatrice
zone *f* de dosage	zona *f* de dosificación; zona *f* de homogeneización	zona *f* dosatrice
course *f* de dosage	carrera *f* de dosificación	corsa *f* di dosaggio
dispositif *m* de dosage	dispositivo *m* de dosificación	dosatore *m*
méthacrylate *m*	metacrilato *m*	metacrilato *m*
méthacrylate *m* de méthyle	metilmetacrilato *m*	metilmetacrilato *m*
mica *m*	mica *f*	mica *f*
micelle *f*	micela *f*	micella *f*
résistance *f* micro-bienne	resistencia *f* a los microorganismos	resistenza *f* ai microbi
caoutchouc *m* microcellulaire	goma *f* porosa	gomma *f* microcellulare
microscopie *f*	microscopía *f*	microscopia *f*
microsphère *f*	microesfera *f*	microsfera *f*
microstructure *f*	microestructura *f*	microstruttura *f*

Nr.	English	Deutsch
949	migration	Migration *f*
950	mill	Mühle *f*
951	milled fibre	gemahlene Faser *f*
952	milling	Fräsen *n*
953	milling machine	Fräsmaschine *f*
954	mineral filler	Mineralfüllstoff *m*
955	miscibility	Mischbarkeit *f*
956	mixer	Mischer *m*
957	mixing head	Mischkopf *m*
958	mixing rolls	Mischwalzwerk *n*
959	mixing section *(screw)*	Mischzone *f*
960	modification	Modifizierung *f*
961	modifier	Modifiziermittel *n*
962	modulus	Modul *m*
963	modulus of elasticity	Elastizitätsmodul *m*
964	moisture content	Feuchtigkeitsgehalt *m*

Français	*Español*	*Italiano*
migration *f*	migración *f*	migrazione *f*
broyeur *m*	molino *m*	frantumatore *m*
fibre *f* broyée	fibra *f* molida	fibra *f* macinata
fraisage *m*	fresado *m*	fresatura *f*
fraiseuse *f*	fresadora *f*	fresatrice *f*
charge *f* minérale	carga *f* mineral	carica *f* minerale
miscibilité *f*	capacidad *f* para ser mezclado	miscibilità *f*
mélangeur *m*	mezclador *m*	mescolatrice *f*
tête *f* de mélange	cabezal *m* mezclador	testa *f* di miscelazione
cylindres *mpl* mélangeurs	mezclador *m* de rodillos	cilindri *mpl* di mescolatura
zone *f* de mélange	zona *f* de mezcla	zona *f* di mescolatura
modification *f*	modificación *f*	modifica *f*
modifiant *m*	modificador *m*	modificante *m*
module *m*	módulo *m*	modulo *m*
module *m* d'élasticité	módulo *m* de elasticidad	modulo *m* di elasticità
taux *m* d'humidité	contenido *m* de humedad	contenuto *m* d'umidità

Nr.	English	Deutsch
965	**moisture streaks**	Feuchtigkeitsschlieren *fpl*
966	**molar mass**	molare Masse *f*
967	**molecular orientation**	Molekülorientierung *f*
968	**molecular sieve**	Molekularsieb *n*
969	**molecular structure**	Molekülstruktur *f*
970	**molecular weight**	Molekulargewicht *n*
971	**molecular weight distribution**	Molekulargewichts-verteilung *f*
972	**monofilament**	Monofilament *n*
973	**monomer**	Monomer *n*
974	**morphology**	Morphologie *f*
975	**mould**	Formwerkzeug *n*
976	**mould breathing**	Werkzeugatmung *f*
977	**mould carrier**	Werkzeugträger *m*
978	**mould changing**	Werkzeugwechsel *m*
979	**mould closing speed**	Schließgeschwindigkeit *f*

Français	*Español*	*Italiano*
traces *fpl* d'humidité	vetas *fpl* de humedad	segni *mpl* di umidità
masse *f* molaire	masa *f* molar	massa *f* molare
orientation *m* moléculaire	orientación *f* molecular	orientazione *f* molecolare
filtre *m* moléculaire	criba *f* molecular	setaccio *m* molecolare
structure *f* moléculaire	estructura *f* molecular	struttura *f* molecolare
poids *m* moléculaire	peso *m* molecular	peso *m* molecolare
distribution *f* de poids moléculaire	distribución *f* de peso molecular	distribuzione *f* del peso molecolare
monofilament *m*	monofilamento *m*	monofilamento *m*
monomère *m*	monómero *m*	monomero *m*
morphologie *f*	morfología *f*	morfologia *f*
moule *m*	molde *m*	stampo *m*
respiration *f* du moule	desgasificación *f* del molde	traspirazione *f* dello stampo
porte-moule *m*	portamolde *m*	base *m* dello stampo
changement *m* de moule	cambio *m* de molde	cambio *m* dello stampo
vitesse *f* de fermeture du moule	velocidad *f* del cierre del molde	velocità *f* di chiusura dello stampo

Nr.	English	Deutsch
980	**mould closing time**	Schließzeit *f*
981	**mould construction**	Formenbau *m;* Werkzeugbau *m*
982	**mould core**	Formkern *m*
983	**mould gap**	Werkzeugspalt *m*
984	**mould insert**	Werkzeugeinsatz *m*
985	**mould land**	Abquetschfläche *f*
986	**mould locking**	Werkzeugzuhaltung *f*
987	**mould opening force**	Werkzeugöffnungskraft *f*
988	**mould parting line**	Werkzeugtrennebene *f*
989	**mould plate**	Werkzeugplatte *f*
990	**mould release agent**	Formtrennmittel *n*
991	**mould temperature**	Werkzeugtemperatur *f*

Français	*Español*	*Italiano*
temps *m* de fermeture du moule	tiempo *m* de cierre del molde	tempo *m* di chiusura dello stampo
fabrication *f* de moule	construcción *f* de moldes	costruzione *f* dello stampo
noyau *m* de moule	núcleo *m* del molde	nucleo *m* dello stampo
espace *m* lumière du moule	espacio *m* entre platos	distanza *f* tra i piani
insert *m* de moule	inserto *m* del molde	inserto *m* dello stampo
surface *f* de contact du moule	superficie *f* de aplastamiento	superficie *f* di contatto di uno stampo
fermeture *f* du moule	fijación *f* del molde	chiusura *f* dello stampo
force *f* d'ouverture du moule	fuerza *f* de apertura del molde	forza *f* di apertura dello stampo
plan *m* de joint du moule	línea *f* de separación del molde	linea *f* di giunzione dello stampo
plateau *m* du moule	placa *f* del molde	piastra *f* dello stampo
agent *m* de démoulage	agente *m* de desmoldeo	agente *m* antiadesivo; distaccante *m*
température *f* de moule	temperatura *f* del molde	temperatura *f* dello stampo

Nr.	English	Deutsch
992	**mould venting**	Werkzeugentlüftung *f*
993	**mould weight**	Werkzeuggewicht *n*
994	**mould, to**	formen
995	**moulding; moulded part**	Formteil *n;* Pressteil *n*
996	**moulding compounds**	Formmasse *f;* Pressmasse *f*
997	**moulding defect** *(inj.)*	Spritzgießfehler *m*
998	**moulding geometry**	Formteilgeometrie *f*
999	**moulding plant**	Spritzgießbetrieb *m*
1000	**moulding pressure**	Pressdruck *m*
1001	**moulding temperature**	Spritzgießtemperatur *f*
1002	**mounting clutch**	Spannpratze *f*
1003	**mounting plate**	Säulenplatte *f;* Buchsenplatte *f*
1004	**mounting support**	Halterung *f*
1005	**moving platen**	Werkzeugaufspannplatte *f,* bewegliche

Français	Español	Italiano
évent *m* de moule	ventilación *f* del molde	degasaggio *m* dello stampo
poids *m* du moule	peso *m* del molde	peso *m* dello stampo
mouler	moldear	stampare
pièce *f* moulée	pieza *f* moldeada	pezzo *m* stampato
compound *m* à mouler	compuestos *mpl* de moldeo	materiali *m* plastici per stampaggio
défaut *m* de moulage	defecto *m* de la inyección	difetto *m* di stampaggio
géométrie *f* de moulage	geometría *f* de la pieza moldeada	geometria *f* della stampata
atelier *m* d'injection	taller *m* de moldeo	fabbrica *f* di stampaggio
pression *f* de moulage	presión *f* de prensado	pressione *f* di stampaggio
température *f* de moulage	temperatura *f* de inyección	temperatura *f* di stampaggio
griffe *f* de montage	garra *f* de montaje	supporto *m* frizione
plaque *f* de montage	placa *f* de fijación	piastra *f* di montaggio
support *m* de montage	soporte *m* de montaje	supporto *m* di montaggio
plateau *f* mobile du moule	placa *f* móvil del molde	piastra *f* mobile

Nr.	English	Deutsch
1006	**multi-axial stress**	Spannung *f*, mehrachsige
1007	**multi-cavity mould**	Mehrfach-Werkzeug *n*
1008	**multi-colour machine**	Mehrfarben-Maschine *f*
1009	**multi-component injection moulding**	Mehrkomponenten-Spritzgießen *n*
1010	**multi-daylight mould**	Etagenwerkzeug *n*
1011	**multi-daylight press; multi-platen press**	Etagenpresse *f*
1012	**multi-layer film**	Verbundfolie *f*; Mehrschichtfolie *f*
1013	**multi-life-feed-injection moulding**	MLFM-Verfahren *n*
1014	**multiple metering**	Mehrfach-Dosierung *f*
1015	**nanocomposite**	Nanocomposite *n*
1016	**nanofiller**	Nanofüllstoff *m*
1017	**nanostructure**	Nanostruktur *f*
1018	**natural resins**	Naturharze *npl*

Français	*Español*	*Italiano*
contrainte *f* multiaxiale	esfuerzo *m* multiaxial	sollecitazione *f* multiassiale
moule *m* multi-empreinte	molde *m* de varias cavidades; molde *m* múltiple	stampo *m* multiplo; stampo *m* a più cavità
machine *f* multi-couleur	máquina *f* para la inyección de varios colores	macchina *f* multicolore
injection *f* multicomposant	moldeo *m* por inyección de varios componentes	stampaggio *m* a iniezione di multicomponenti
moule *m* à étages	molde *m* de pisos	stampo *m* multipiani
presse *f* à étages, multiplateaux	prensa *f* de platos múltiples	pressa *f* a più aperture
film *m* multicouche	película *f* multicapa	film *m* multistrato
moulage *m* multi-injection	proceso *m* MLFM	stampaggio *m* ad alimentazione multipla
dosage *m* multiple	dosificación *f* múltiple	dosaggio *m* multiplo
nanocomposite *m*	nanocomposite *m*	nanocomposito *m*
nanocharge *f*	nanocarga *f*	nanocarica *f*
nanostructure *f*	nanoestructura *f*	nanostruttura *f*
résines *fpl* naturelles	resinas *fpl* naturales	resine *fpl* naturali

Nr.	English	Deutsch
1019	**natural rubber**	Naturkautschuk *m*
1020	**necking**	Neck-Bildung *f;* Halsbildung *f*
1021	**needle valve**	Nadelventil *n;* Nadelverschluss *m*
1022	**needle valve nozzle**	Nadelverschlussdüse *f*
1023	**nip between rolls**	Walzenspalt *m*
1024	**nip rolls**	Abquetschwalzen *fpl*
1025	**nitrided steel**	Nitrierstahl *m*
1026	**nitrile rubber**	Nitrilkautschuk *m*
1027	**non-destructive testing**	zerstörungsfreie Prüfung *f*
1028	**non-return valve**	Rückstromsperre *f*
1029	**non-woven**	Vlies *n;* Vliesstoff *m*
1030	**notched**	gekerbt
1031	**notched impact strength**	Kerbschlagzähigkeit *f*
1032	**nozzle**	Düse *f*

Français	Español	Italiano
caoutchouc *m* naturel	caucho *m* natural	gomma *f* naturale
necking *m*	estrechamiento *m*	strozzatura *f*
clapet *m* à aiguilles	válvula *f* de aguja	valvola *f* a spillo
obturateur *m* à aiguilles	boquilla *f* con obturador	ugello *m* con valvola a spillo
espace *m* entre cylindres	abertura *f* entre cilindros	scarto *m* fra rulli
rouleaux *mpl* pinceurs	rodillos *mpl* de compresión	cilindri *mpl* di traino
acier *f* nitruré	acero *m* nitrurado	acciaio *m* nitrurato
caoutchouc *m* nitrile	caucho *m* nitrílico	gomma *f* nitrilica
contrôle *m* non destructif	prueba *f* no destructiva	prova *f* non distruttiva
clapet *m* anti retour	válvula *f* antirretorno	valvola *f* di non ritorno
nontissé *m*	fieltro *m;* tejido *m* no tejido	tessuto *m* non tessuto
entaillé	entallado	con intaglio
résistance *f* au choc avec entaille	resistencia *f* al impacto con entalla	resistenza *f* all'urto con intaglio
buse *f*	boquilla *f;* tobera *f*	ugello *m*

Nr.	English	Deutsch
1033	**nozzle heater**	Düsenheizung *f*
1034	**nozzle point**	Düsenspitze *f*
1035	**nucleating agent**	Nukleierungsmittel *n;* Keimbildner *m*
1036	**nucleation**	Keimbildung *f*
1037	**number of cavities**	Formnestzahl *f;* Fachzahl *f*
1038	**number of cycles**	Taktzahl *f*
1039	**number of flights** *(screw)*	Gangzahl *f*
1040	**O-ring seal**	O-Ring-Dichtung *f*
1041	**offset printing**	Offsetdruck *m*
1042	**oil resistance**	Ölbeständigkeit *f*
1043	**olefins**	Olefine *npl*
1044	**oligomer**	Oligomer *n*
1045	**opaque**	opak
1046	**open-cell** *(foam)*	offenzellig
1047	**opening force**	Öffnungskraft *f*

Français	Español	Italiano
réchauffeur *m* de buse	calefactor *m* de la boquilla	riscaldatore *m* dell'ugello
tête *f* de la buse	punta *f* de la boquilla	puntale *m* ugello
agent *m* nucléant	agente *m* nucleante	agente *m* nucleante
nucléation *f*	nucleación *f*	nucleazione *f*
nombre *m* d'empreintes	número *m* de cavidades	numero *m* di cavità
nombre *m* de cycles	número *m* de ciclos	numero *m* di cicli
nombre *m* de filets	número *m* de filetes	numero *m* di filetti
joint *m* torique	retén *m* de anillo	guarnizione *f* ad anello
impression *f* offset	impresión *f* offset	stampa *m* offset
résistance *f* aux huiles	resistencia *f* al aceite	resistenza *f* all'olio
oléfines *fpl*	olefinas *fpl*	olefine *fpl*
oligomère *m*	oligómero *m*	oligomero *m*
opaque	opaco	opaco
à cellules ouvertes (mousse)	célula *f* abierta	a celle aperte (espanso)
force *f* d'ouverture	fuerza *f* de apertura	forza *f* di apertura

Nr.	English	Deutsch
1048	**opening stroke**	Öffnungsweg *m*
1049	**operating point**	Arbeitspunkt *m*
1050	**optical brightener**	optischer Aufheller *m*
1051	**optical properties**	optische Eigenschaften *fpl*
1052	**orange peel effect**	Apfelsinenschalenhaut-effekt *m*
1053	**organic**	organisch
1054	**orientation**	Orientierung *f*
1055	**output**	Ausstoß *m*
1056	**output rate**	Durchsatzleistung *f*
1057	**outsert moulding**	Outsert-Technik *f*
1058	**overcuring**	Überhärtung *f*
1059	**overheating**	Überhitzen *n*
1060	**overpacking**	Überspritzen *n*
1061	**oxidation**	Oxidation *f*
1062	**ozone resistance**	Ozonbeständigkeit *f*

Français	*Español*	*Italiano*
course *f* d'ouverture	recorrido *m* de la abertura	corsa *f* di apertura
conditions *fpl* opératoires	punto *m* de operaciones	punto *m* operativo
azurant *m* optique	blanqueante *m* óptico	sbiancante *m* ottico
propriétés *fpl* optiques	propiedades *fpl* ópticas	proprietà *fpl* ottiche
effet *m* de peau d'orange	efecto *m* piel de naranja	effetto *m* buccia d'arancia
organique	orgánico	organico
orientation *f*	orientación *f*	orientamento *m*
débit *m*	producción *f*	produzione *f*
niveau *m* de débit	rendimiento *m* de producción	livello *m* di produzione
outsert *m* moulding	técnica *f* «outsert»	stampaggio *m* „outsert"
surcuisson *f*	sobrecurado *m*	sovratrattamento *m*
surchauffe *f*	sobrecalentamiento *m*	surriscaldamento *m*
bourrage *m*	sobreinyección *m*	sovraimballaggio *m*
oxydation *f*	oxidación *f*	ossidazione *f*
résistance *f* à l'ozone	resistencia *f* al ozono	resistenza *f* all'ozono

Nr.	English	Deutsch
1063	**packing piston**	Nachpresskolben *m*
1064	**paddle mixer**	Schaufelrührer *m*
1065	**paint**	Farbe *f;* Lack *m*
1066	**parison**	Vorformling *m*
1067	**parison die**	Schlauchkopf *m*
1068	**partial wall thickness control**	partielle Wanddickensteuerung (PWDS) *f*
1069	**particle**	Partikel *f*
1070	**particle foam**	Partikelschaumstoff *m*
1071	**particle size**	Teilchengröße *f*
1072	**parting line** *(mould)*	Trennebene *f*
1073	**paste**	Paste *f*
1074	**peel strength**	Schälfestigkeit *f*
1075	**peel test**	Schälversuch *m*
1076	**peeling machine**	Schälmaschine *f*
1077	**pellet**	Granulat *n*

Français	*Español*	*Italiano*
piston *m* de compactage	pistón *m* de presión posterior	pistone *m* di compattazione
mélangeur *m* à palettes	turboagitador *m*	agitatore *m* a palette
peinture *f*	laca *f;* pintura *f*	lacca *f;* vernice *f*
paraison *f*	preforma *f;* parisón *m*	parison *m;* preforma *f*
tête *f* de paraison	boquilla *f* de parisón	filiera *f* del parison
régulation *f* partielle d'épaisseur de paroi	regulación *f* parcial del espesor de pared	controllo *m* parziale dello spessore di parete
particule *f*	partícula *f*	particella *f*
mousse *f* de particules	partícula *f* de espuma	espanso *m* a granuli
taille *f* de particule	tamaño *m* de partícula	granulometria *f*
plan *m* de joint	linea *f* de separación	linea *f* di giunzione
pâte *f*	pasta *f*	pasta *f*
résistance *f* au pelage	resistencia *f* al pelado	resistenza *f* alla defoliazione
essai *m* de pelage	ensayo *m* de pelado	prova *f* di defoliazione
machine *f* de pelage	máquina *f* peladora	sfogliatrice *f*
granulé *m*	granulado *m*	granulo *m*

Nr.	English	Deutsch
1078	**pelletiser**	Granuliervorrichtung *f*
1079	**pelletise, to**	granulieren
1080	**pendulum impact test**	Pendelschlagversuch *m*
1081	**penetration test**	Durchstoßversuch *m*
1082	**perforated plate**	Lochplatte *f*
1083	**perforation**	Perforation *f*
1084	**permeability**	Durchlässigkeit *f*
1085	**permeation**	Permeation *f*
1086	**peroxide**	Peroxid *n*
1087	**petrochemicals**	Petrochemikalien *fpl*
1088	**phase separation**	Phasentrennung *f*
1089	**phase transition**	Phasenübergang *m*
1090	**phenol formaldehyde resin** *(PF)*	Phenol-Formaldehyd-Harz *n*
1091	**phenolic foam**	Phenolharz-Schaumstoff *m*
1092	**phenolic moulding compound**	Phenolharz-Formmasse *f*

Français	*Español*	*Italiano*
granulateur *m*	granulador *m*	estrusore *m* granulare
granuler	granular	granulare
test *m* au choc du pendule	ensayo *m* de impacto con péndulo	prova *f* d'urto con pendolo
test *m* de perforation	ensayo *m* de penetración	prova *f* di penetrazione
plaque *f* perforée	placa *f* perforada	piastra *f* perforata
perforation *f*	perforación *f*	perforazione *f*
perméabilité *f*	permeabilidad *f*	permeabilità *f*
perméation *f*	permeabilidad *f*	permeazione *f*
peroxyde *m*	peróxido *m*	perossido *m*
produits *mpl* pétrochimiques	productos *mpl* petroquímicos	petrolchimici *mpl*
séparation *f* de phase	separación *f* de fases	separazione *f* di fase
transition *f* de phase	transición *f* de fases	transizione *f* delle fasi
résine *f* phénol formaldéhyde	resina *f* de fenol-formaldehido	resina *f* fenol formaldeide
mousse *f* phénolique	espuma *f* fenólica	espanso *m* fenolico
matière *f* à mouler phénolique	compuesto *m* fenólico para moldeo	mescola *f* per stampaggio fenolica

Nr.	English	Deutsch
1093	**phenolic resins**	Phenolharze *npl*
1094	**photooxidation**	Photooxidation *f*
1095	**photoresist**	Photoresist *m*
1096	**pigment**	Pigment *n*
1097	**pilot plant**	Pilotanlage *f*
1098	**pin**	Stift *m*
1099	**pin gate**	Punktanguss *m*
1100	**pin-lined barrel extruder**	Stiftzylinder-Extruder *m*
1101	**pinch-off area**	Quetschzone *f*
1102	**pinch-off weld flash**	Quetschnahtwulst *m*
1103	**pipe**	Rohr *n*
1104	**piston**	Kolben *m*
1105	**planetary-gear extruder**	Planetwalzenextruder *m*
1106	**plant**	Anlage *f;* Produktionsstätte *f*

Français	*Español*	*Italiano*
résines *fpl* phénoliques	resinas *fpl* fenólicas	resine *fpl* fenoliche
photo-oxydation *f*	fotooxidación *f*	fotoossidazione *f*
photorésist *m*	fotorresistente *m*	fotoresistente *m*
pigment *m*	pigmento *m*	pigmento *m*
unité *f* pilote	instalación *f* piloto	impianto *m* pilota
aiguille *f*	espiga *f*	perno *m*
injection *f* capillaire	entrada *m* capilar	iniezione *f* capillare
cylindre *m* d'extrudeuse à picots	extrusora *f* de cilindro cónico	estrusore *m* con cilindro a pioli
zone *f* de pinçage	zona *f* de aplastamiento; zona *f* de soldadura	area *f* di pinzatura
bourrelet *m* de pinçage	cordón *m* de la costura de aplastamiento	bava *f* di saldatura di pinzatura
tuyau *m;* tube *m*	tubo *m*	tubo *m*
piston *m*	pistón *m*	pistone *m*
extrudeuse *f* planétaire	extrusora *f* de rodillos planetarios	estrusore *m* a viti planetarie
usine *f;* unité *f*	planta *f* de producción	impianto *m;* fabbrica *f*

Nr.	English	Deutsch
1107	**plasma polymerisation**	Plasmapolymerisation *f*
1108	**plasma treatment**	Plasmabehandlung *f*
1109	**plastic foam;** **foam plastic**	Schaumkunststoff *m*
1110	**plasticise, to**	plastifizieren
1111	**plasticiser**	Weichmacher *m*
1112	**plasticising capacity**	Plastifizierleistung *f*
1113	**plasticising cylinder**	Plastifizierzylinder *m*
1114	**plasticising section** *(of a screw)*	Plastifizierzone *f*
1115	**plasticity**	Plastizität *f*
1116	**plastics**	Kunststoffe *mpl*
1117	**plastics processing**	Kunststoffverarbeitung *f*
1118	**plastification;** **plastication**	Plastifizierung *f*
1119	**plastificator; plasticator**	Plastifikator *m*
1120	**plastisol**	Plastisol *n*

Français	*Español*	*Italiano*
polymérisation *f* par plasma	polimerización *f* por plasma	polimerizzazione *f* al plasma
traitement *m* au plasma	tratamiento *m* por plasma	trattamento *m* al plasma
mousse *f* plastique	espuma *f* plástica	espanso *m* plastico
plastifier	plastificar	plastificare
plastifiant *m*	plastificante *m*	plastificante *m*
capacité *f* de plastification	capacidad *f* de plastificación	capacità *f* di plastificazione
cylindre *m* de plastification	cilindro *m* de plastificación	cilindro *m* di plastificazione
zone *f* de plastification	zona *f* de plastificación	zona *f* di plastificazione
plasticité *f*	plasticidad *f*	plasticità *f*
matières *fpl* plastiques; plastiques *mpl*	plásticos *mpl;* materias *fpl* plásticas	materie *fpl* plastiche
transformation *f* des plastiques	transformación *f* de plásticos	trasformazione *f* delle materie plastiche
plastification *f*	plastificación *f*	plastificazione *f*
plastificateur *m*	plastificador *m*	plastificante *m*
plastisol *m*	plastisol *m*	plastisol *m*

Nr.	English	Deutsch
1121	**plate-out**	Belagbildung *f*
1122	**platen**	Werkzeugaufspannplatte *f*
1123	**platen area**	Aufspannfläche *f*
1124	**platen press**	Plattenpresse *f*
1125	**plunger**	Kolben *m*
1126	**plunger injection**	Kolbeneinspritzung *f*
1127	**plunger injection moulding maschine**	Kolbenspritzgießmaschine *f*
1128	**pneumatic**	pneumatisch
1129	**pneumatic feeder**	pneumatische Förderung *f*
1130	**Poisson's ratio**	Poisson-Zahl *f;* Querkontraktionszahl *f*
1131	**polishing**	Polieren *n*
1132	**polishing calender**	Glättkalander *m*
1133	**polishing nip**	Glättspalt *m*
1134	**polishing rolls**	Glättwalzen *fpl*

Français	Español	Italiano
plate-out	deposición *m*	deposito *m*
plateau *m* porte-moule	placa *f* portamolde	piastra *f;* piano *m* di pressa
surface *f* de plateau	superficie *f* de la placa portamoldes	area *f* della piastra
presse *f* à plateaux	prensa *f* de platos	pressa *f* a piani
piston *m*	pistón *m;* émbolo *m*	pistone *m*
injection *f* par piston	inyección *f* mediante pistón	iniezione *f* a pistone
machine *f* d'injection à piston	máquina *f* de inyección por pistón	pressa *f* per iniezione a trasferimento con pistone ausiliario
pneumatique	neumático	pneumatico
alimentation *f* pneumatique	transporte *m* neumático	trasportatore *m* pneumatico
coefficient *m* de Poisson	relación *f* de Poisson	coefficiente *m* di Poisson
polissage *m*	pulimento *m*	lucidatura *f*
calandre *f* de polissage	calandra *f* de pulido	calandra *f* di lucidatura
espace *m* entre deux rouleaux	espacio *m* de pulido entre rodillos	passaggio *m* di lucidatura
rouleaux *mpl* de polissage	rodillos *mpl* de satinado	rulli *mpl* per satinare

Nr.	English	Deutsch
1135	**polishing stack**	Glättwerk *n*
1136	**polishing wheel**	Polierscheibe *f*
1137	**polyacetal** *(POM)*	Polyacetal *n*
1138	**polyacrylonitrile** *(PAN)*	Polyacrylnitril *n*
1139	**polyaddition**	Polyaddition *f*
1140	**polyamide** *(PA)*	Polyamid *n*
1141	**polyamide imide** *(PAI)*	Polyamidimid *n*
1142	**polybutylene terephthalate**	Polybutylenterephthalat *n*
1143	**polycarbonate** *(PC)*	Polycarbonat *n*
1144	**polycondensation**	Polykondensation *f*
1145	**polyelectrolytes**	Polyelektrolyte *npl*
1146	**polyester**	Polyester *m*
1147	**polyether**	Polyether *m*
1148	**polyether ether ketone** *(PEEK)*	Polyetheretherketon *n*
1149	**polyether imide** *(PEI)*	Polyetherimid *n*
1150	**polyether sulphone** *(PES)*	Polyethersulfon *n*
1151	**polyethylene** *(PE)*	Polyethylen *n*

Français	*Español*	*Italiano*
meule *f* de polissage	cilindro *m* de pulido	rullo *m* di lucidatura
roue *f* de polissage	disco *m* de pulido	disco *m* pulitore
polyacétal *m*	poliacetal *m*	poliacetale *m*
polyacrylonitrile *m*	poliacrilonitrilo *m*	poliacrilonitrile *m*
polyaddition *f*	poliadición *f*	poliaddizione *f*
polyamide *m*	poliamida *f*	poliamide *m*
polyamide-imide *m*	poliamidimida *f*	poliamideimide *m*
polybutylène-téréphtalate *m*	politereftalato *m* de butilo	polibutilentereftalato *m*
polycarbonate *m*	policarbonato *m*	policarbonato *m*
polycondensation *f*	policondensación *f*	policondensazione *f*
polyélectrolytes *mpl*	polielectrolitos *mpl*	polielettroliti *mpl*
polyester *m*	poliéster *m*	poliestere *m*
polyéther *m*	poliéter *m*	polietere *m*
polyétheréthercétone *m*	polieterétercetona *f*	polietere-etere-chetone *m*
polyétherimide *m*	poliéterimida *f*	polietereimmide *f*
polyéthersulfone *m*	poliétersulfona *f*	polieteresulfone *m*
polyéthylène *m*	polietileno *m*	polietilene *m*

Nr.	English	Deutsch
1152	**polyethylene terephthalate** *(PET)*	Polyethylenterephthalat *n*
1153	**polyisocyanate**	Polyisocyanat *n*
1154	**polymer alloy**	Kunststoff-Legierung *f*
1155	**polymer blend**	Kunststoff-Mischung *f;* Polymer-Blend *n*
1156	**polymer concrete**	Polymerbeton *m;* Reaktionsharzbeton *m*
1157	**polymer dispersion**	Polymer-Dispersion *f*
1158	**polymer homologues**	Polymerhomologe *npl*
1159	**polymerisation**	Polymerisation *f*
1160	**polymerisation aid**	Polymerisations-Hilfsmittel *n*
1161	**polymers**	Polymere *npl*
1162	**polymethyl methacrylate** *(PMMA)*	Polymethylmethacrylat *n*
1163	**polyolefins** *(PO)*	Polyolefine *npl*
1164	**polyols**	Polyole *npl*
1165	**polyphenylene ether** *(PPE),* **modified**	Polyphenylenether *m,* modifiziert

Français	*Español*	*Italiano*
polyéthylène téréphtalate *m*	politereftalato de etileno *m*	polietilentereftalato *m*
polyisocyanate *m*	poliisocianato *m*	poliisocianato *m*
alliage *m* de polymères	aleación *f* de polímeros	lega *f* polimerica
blend *m* de polymères	mezclas *fpl* de polímeros	mescola *f* polimerica
béton *m* de résine	hormigón *m* de resinas de reacción; cemento *m* polimérico	cemento *m* polimerico
dispersion *f* de polymère	dispersión *f* de polímeros	dispersione *f* polimerica
homologues *mpl* de polymères	polímeros *mpl* homólogos	omologhi *mpl* polimerici
polymérisation *f*	polimerización *f*	polimerizzazione *f*
agent *m* de polymérisation	agente *m* de polimerización	agente *m* polimerizzante
polymères *mpl*	polímeros *mpl*	polimeri *mpl*
polyméthacrylate *m* de méthyle	polimetacrilato *m* de metilo	polimetilmetacrilato *m*
polyoléfines *fpl*	poliolefinas *fpl*	poliolefine *fpl*
polyols *mpl*	polioles *mpl*	polioli *mpl*
polyéther *m* de phénylène, modifié	poliéter *m* de fenileno, modificado	polifenilenetere *m*, modificato

Nr.	English	Deutsch
1166	**polyphenylene sulphide** (PPS)	Polyphenylensulfid *n*
1167	**polypropylene** *(PP)*	Polypropylen *n*
1168	**polystyrene** *(PS)*	Polystyrol *n*
1169	**polysulphone** *(PSU)*	Polysulfon *n*
1170	**polytetrafluoroethylene** *(PTFE)*	Polytetrafluorethylen *n*
1171	**polyurethane** *(PU)*	Polyurethan *n*
1172	**polyurethane elastomers**	Polyurethan-Elastomere *npl*
1173	**polyurethane foam**	Polyurethan-Schaumstoff *m*
1174	**polyurethane rubber**	Polyurethan-Kautschuk *m*
1175	**polyvinyl chloride** (PVC)	Polyvinylchlorid *n*
1176	**post-cooling**	Nachkühlung *f*
1177	**post-cooling mould**	Nachkühlwerkzeug *n*
1178	**post-crystallisation**	Nachkristallisation *f*
1179	**post-curing**	Nachhärten *n*
1180	**post-forming**	Nachverformung *f*

Français	*Español*	*Italiano*
polysulfure *m* de phénylène	polisulfuro *m* de fenileno	solfuro *m* di polifenilene
polypropylène *m*	polipropileno *m*	polipropilene *m*
polystyrène *m*	poliestireno *m*	polistirene *m*
polysulfone *m*	polisulfona *f*	polisolfone *m*
polytétrafluoro-éthylène *m*	politetrafluor-etileno *m*	politetrafluoroetilene *m*
polyuréthanne *m*	poliuretano *m*	poliuretano *m*
élastomères *mpl* de polyuréthanne	elastómeros *mpl* de poliuretano	elastomeri *mpl* poliuretanici
mousse *f* polyuréthanne	espuma *f* de poliuretano	espanso *m* poliuretanico
caoutchouc *m* de polyuréthanne	caucho *m* de poliuretano	gomma *f* poliuretanica
chlorure *m* de polyvinyle	policloruro *m* de vinilo	cloruro *m* di polivinile
post-refroidissement *m*	post-enfriamiento *m*	post-raffreddamento *m*
moule *m* de post refroidissement	molde *m* de postenfriamiento	stampo *m* a postraffreddamento
post-cristallisation *f*	cristalización *f* posterior	postcristallizzazione *f*
post-cuisson *f*	postcurado *m*	post-trattamento *m*
post formage *m*	postconformado *m*	postformatura *f*

Nr.	English	Deutsch
1181	**post-shrinkage**	Nachschwindung *f*
1182	**pot life**	Topfzeit *f*
1183	**powder coating**	Pulverlackbeschichtung *f*
1184	**powder injection moulding**	Pulverspritzguss *m*
1185	**pre-drying**	Vortrocknen *n*
1186	**pre-heating**	Vorwärmen *n*
1187	**pre-plasticising**	Vorplastifizieren *n*
1188	**precipitation polymerisation**	Fällungspolymerisation *f*
1189	**precision injection moulding**	Präzisionsspritzgießen *n*
1190	**preconditioning**	Vorbehandlung *f*
1191	**preform**	Vorformling *m*
1192	**premix**	Vormischung *f*
1193	**prepolymer**	Präpolymer *n*
1194	**prepreg**	Prepreg *m;* Harzmatte *f*
1195	**press**	Presse *f*

Français	*Español*	*Italiano*
post-retrait *m*	contracción *f* posterior al moldeo	post-ritiro *m*
pot life *m*, durée de vie *f* en pot	vida *f* útil de aplicación	durata *f* di lavorabilità
revêtement *m* par poudre	recubrimiento *m* a partir de polvo	rivestimento *m* con polvere
injection *f* de poudre	moldeo *m* por inyección de polvo	stampaggio *m* a iniezione di polveri
pré-séchage *m*	presecado *m*	pre-essiccazione *f*
préchauffage *m*	precalentamiento *m*	preriscaldamento *m*
pré-plastification *f*	preplastificación *f*	preplastificazione *f*
polymérisation *f* par précipitation	polimerización *f* por precipitación	polimerizzazione *f* per precipitazione
injection *f* de précision	inyección *m* de precisión	stampaggio *m* a iniezione di precisione
préconditionne-ment *m*	preacondiciona-miento *m*	precondizionamento *m;* pretrattamento *m*
préforme *f*	preforma *f*	preforma *f*
premix *m*	premezcla *f*	premiscela *f*
prépolymère *m*	prepolímero *m*	prepolimero *m*
préimprégné *m*	preimpregnado *m*	preimpregnato *m*
presse *f*	prensa *f*	pressa *f*

Nr.	English	Deutsch
1196	pressure accumulator	Druckspeicher *m*
1197	pressure control valve	Druckregelventil *n*
1198	pressure feeder	Druckförderer *m*
1199	pressure flow	Druckströmung *f*
1200	pressure gelation process	Druckgelierverfahren *n*
1201	pressure loss	Druckverlust *m*
1202	pressure roll	Anpresswalze *f*
1203	pretreatment	Vorbehandlung *f*
1204	primer	Primer *m*
1205	printing	Bedrucken *n*
1206	printing ink	Druckfarbe *f*
1207	process control	Prozesssteuerung *f*
1208	processing	Verarbeitung *f*

Français	*Español*	*Italiano*
accumulateur *m* de pression	acumulador *m* de presión	accumulatore *m* di pressione
vanne *f* de contrôle de pression	válvula *f* de regulación de la presión	valvola *f* per controllo pressione
système *f* d'alimentation par pression	transporte *m* por presión	trasportatore *m* per pressione
flux *m* de pression	flujo *m* de presión	flusso *m* di pressione
procédé *m* de gélification par pression	proceso *m* de gelificación a presión	processo *m* di gelificazione a pressione
perte *f* de pression	pérdida *f* de presión	perdita *f* di pressione
cylindre *m* de pression	rodillo *m* de presión	rullo *m* di pressione
préconditionnement *m*	pretratamiento *m*	precondizionamento *m;* pretrattamento *m*
primer *m*	capa *f* de preparación	mano *f* di fondo
impression *f*	impresión *f*	stampa *f*
encre *f* d'impression	tinta *f* de impresión	inchiostro *m* da stampa
contrôle *m* de processus	control *m* de proceso	controllo *m* di processo
transformation *f*	transformación *f*	lavorazione *f;* trasformazione *f*

Nr.	English	Deutsch
1209	**processing aids**	Verarbeitungshilfsmittel *npl*
1210	**processing temperature**	Verarbeitungstemperatur *f*
1211	**processor**	Verarbeiter *m*
1212	**production**	Produktion *f*
1213	**production speed**	Produktionsgeschwindigkeit *f*
1214	**productivity**	Produktivität *f*
1215	**profile**	Profil *n*
1216	**projected area**	projizierte Fläche *f*
1217	**property**	Eigenschaft *f*
1218	**proportional valve**	Proportionalventil *n*
1219	**protective colloids**	Schutzkolloide *npl*
1220	**prototype**	Prototyp *m*
1221	**prototype mould**	Prototypenwerkzeug *n*
1222	**pseudoplasticity**	Pseudoplastizität *f*
1223	**pultrusion**	Pultrusion *f*

Français	Español	Italiano
agents *mpl* d'aide de la transformation	agentes *mpl* auxiliares para la transformación	ausiliari *mpl* di lavorazione
température *f* de transformation	temperatura *f* de transformación	temperatura *f* di lavorazione
transformateur *m*	transformador *m*	trasformatore *m*
production *f*	producción *f*	produzione *f*
vitesse *f* de production	velocidad *f* de producción	velocità *f* di produzione
productivité *f*	productividad *f*	produttività *f*
profilé *m*	perfil *m*	profilo *m*
surface *f* de projection	superficie proyectada	area *f* proiettata
propriété *f*	propiedad *f*	proprietà *f*
vanne *f* proportionnelle	válvula *f* proporcional	valvola *f* proporzionale
colloïdes *mpl* protecteurs	coloides *mpl* protectores	colloidi *mpl* protettivi
prototype *m*	prototipo *m*	prototipo *m*
moule *m* prototype	molde *m* prototipo	stampo *m* prototipo
pseudoplasticité *f*	seudoplasticidad *f*	pseudoelasticità *f*
pultrusion *f*	pultrusión *f*	pultrusione *f*

Nr.	English	Deutsch
1224	**pulverisation**	Pulverisierung *f*
1225	**punch indentation test**	Stempeleindruckversuch *m*
1226	**punching**	Stanzen *n*
1227	**purging compound**	Reinigungsgranulat *n*
1228	**push-pull injection moulding**	Gegentaktspritzgießen *n*
1229	**pyrolysis**	Pyrolyse *f*
1230	**quality**	Qualität *f*
1231	**quality assurance**	Qualitätssicherung *f*
1232	**quenching**	Abschrecken *n*
1233	**quick-change device**	Schnellwechselvorrichtung *f*
1234	**quick-clamping device**	Schnellspannvorrichtung *f*
1235	**quick-connect coupling**	Schnellkupplung *f*
1236	**quick mould change**	Werkzeugschnellwechsel *m*

Français	*Español*	*Italiano*
pulvérisation *f*	pulverización *f*	polverizzazione *f*
test *m* d'écrasement	ensayo *m* de huella de punzón	prova *f* di penetrazione con punzone
estampage *m*	punzonado *m*	punzonatura *f*
produit *m* de purge	granulado *m* de limpieza	granulato *m* di pulitura
injection *f* push-pull	inyección *f* push-pull	stampaggio *m* a iniezione push-pull (spingi e tira)
pyrolyse *f*	pirólisis *f*	pirolisi *f*
qualité *f*	calidad *f*	qualità *f*
assurance *f* qualité	aseguramiento *m* de la calidad	assisurazione *f* di qualità
piégeage *m* des radicaux	enfriamineto *m* brusco	raffreddamento *m* rapido
dispositif *m* de changement rapide	mecanismo *m* de cambio rápido	dispositivo *m* di cambio rapido
dispositif *m* de couplage rapide	dispositivo *m* de cierre rápido	dispositivo *m* a bloccaggio rapido
bridage *m* rapide	acoplamiento *m* de rápida conexión	accoppiamento *m* a inserimento rapido
changement *m* rapide de moule	cambio *m* rápido de molde	cambio *m* stampi rapido

Nr.	English	Deutsch
1237	**radial screw clearance**	Schneckenspiel *n*
1238	**radiation crosslinking**	Strahlenvernetzung *f*
1239	**radiation curing**	Strahlungshärtung *f*
1240	**radiation resistance**	Strahlenbeständigkeit *f*
1241	**radical**	Radikal *n*
1242	**radical polymerisation**	radikalische Polymerisation *f*
1243	**rake angle**	Spanwinkel *m*
1244	**ram extruder**	Kolbenextruder *m;* Ramextruder *m*
1245	**rapid prototyping**	Rapid Prototyping *n*
1246	**rate of deformation**	Deformationsgeschwindigkeit *f*
1247	**rate of heating**	Aufheizrate *f*
1248	**raw material**	Rohstoff *m*
1249	**reaction injection moulding** *(RIM)*	Reaktionsspritzgießen *n*

Français	Español	Italiano
pas radial *m* de vis	juego *m* radial del tornillo	apertura *f* a vite radiale
réticulation *f* par radiation	reticulación *f* por radiaciones	reticolazione *f* per radiazione
cuisson *m* par radiation	curado *m* por radiación	trattamento *m* per radiazione
résistance *f* au rayonnement	resistencia *f* a radiaciones	resistenza *f* alla radiazione
radical *m*	radical *m*	radicale *m*
polymérisation *f* radicalaire	polimerización *f* radical	polimerizzazione *f* radicalica
angle *m* de râcle; angle *m* de coupe	ángulo *m* de corte	angolo *m* di taglio
extrudeuse *f* à piston	extrusora *f* de pistón	estrusore *m* a pistone
prototypage *m* rapide	generación *f* rápida de prototipos	prototipazione *f* rapida
vitesse *f* de déformation	velocidad *f* de deformación	velocità *f* di deformazione
taux *m* de chauffage	factor *m* de calefacción	tasso *m* di riscaldamento
matière *f* première	materia *f* prima	materia *f* prima
moulage *m* par injection-réaction	moldeo *m* por inyección-reacción	stampaggio *m* per reazione iniezione

Nr.	English	Deutsch
1250	**reaction kinetics**	Reaktionskinetik f
1251	**reaction rate**	Reaktionsgeschwindigkeit f
1252	**reactive extrusion**	reaktive Extrusion f
1253	**reactive resin**	Reaktionsharz n
1254	**reactive solvent/diluent**	reaktives Lösemittel/ Verdünnungsmittel n
1255	**rebound test**	Rückprallversuch m
1256	**reciprocating screw**	Schneckenkolben m
1257	**reciprocating-screw injection**	Schneckenkolben-einspritzung f
1258	**reclaiming**	Wiederverwertung f
1259	**recovery**	Rückverformung f
1260	**recycle, to**	rezyklieren
1261	**recycling of waste**	Abfallverwertung f
1262	**reducing flange**	Reduzierstück n
1263	**reel**	Wickel m
1264	**reel changing**	Rollenwechsel m

Français	*Español*	*Italiano*
cinétique *f* de réaction	cinética *f* química	cinetica *f* di reazione
vitesse *f* de réaction	velocidad *f* de reacción	velocità *f* di reazione
extrusion *f* réactive	extrusión *f* reactiva	estrusione *f* reattiva
résine *f* réactive	resina *f* de reacción	resina *f* reattiva
solvant/diluant *m* réactif	disolvente/diluyente *m* reactivo	solvente/diluente *m* reattivo
test *m* de rebond	ensayo *m* de elasticidad de rebote	prova *f* di ricaduta
vis-piston *f*	émbolo *m* de tornillo	vite *f* punzonante
injection *f* avec piston	inyección *f* por tornillo-émbolo	iniezione *f* con vite punzonante
recupération *f*	recuperación *f*	ricupero *m*
reprise *f* élastique	memoria *f* elástica	memoria *f* elastica
recycler	reciclar	riciclare
recyclage *m* des déchets	reutilización *m* de desperdicios	riciclaggio *m* di rifiuti; sfridi *mpl*
bride *f* de réduction	reductor *m*	flangia *m* riduttrice
bobine *f*	bobina *f;* rollo *m*	bobina *f*
changement *m* de bobine	cambio *m* de bobinas	cambio *m* di bobina

Nr.	English	Deutsch
1265	**reference materials**	Referenzmaterialien *npl*
1266	**reflection coefficient**	Reflexionsgrad *m*
1267	**refractive index**	Brechungsindex *m;* Brechungszahl *f*
1268	**regenerate, to**	regenerieren
1269	**regrind**	Mahlgut *n;* Regenerat *n*
1270	**reinforced plastics**	verstärkte Kunststoffe *mpl*
1271	**reinforcement**	Verstärkung *f*
1272	**reinforcing agent**	Verstärkungsmittel *n*
1273	**rejects**	Ausschuss *m*
1274	**relaxation**	Relaxation *f*
1275	**relaxation time**	Relaxationszeit *f*
1276	**release agent**	Trennmittel *n*
1277	**release coating**	Release-Schicht *f*
1278	**release film**	Trennfolie *f*
1279	**repair**	Reparatur *f*

Français	*Español*	*Italiano*
matériaux *mpl* de référence	materiales *mpl* de referencia	materiali *mpl* di riferimento
coefficient *m* de réflection	coeficiente *m* de reflexión	coefficiente *m* di riflessione
indice *m* de réfraction	índice *m* de refracción	indice *m* di rifrazione
régénérer	regenerar	rigenerare
rebroyé *f*	molienda *f*	rimacinato *m;* materiale *m* di ricupero
plastiques *mpl* renforcés	plásticos *mpl* reforzados	materie *fpl* plastiche rinforzate
renforcement *m*	refuerzo *m*	rinforzo *m*
agent *m* de renfort	agente *m* de refuerzo	agente *m* di rinforzo
rebut *m*	desperdicio *m*	scarti *mpl*
relaxation *f*	relajación *f*	rilassamento *m*
temps *m* de relaxation	tiempo *m* de relajación	tempo *m* di rilassamento
agent *m* de démoulage	agente *m* desmoldeante	agente *m* di separazione
couche *f* antiadhérente	capa *f* desmoldeante	agente *m* di distacco
feuille *f* de séparation	hoja *f* de separación	film *m* di distacco
réparation *f*	reparación *f*	riparazione *f*

Nr.	English	Deutsch
1280	**reprocessing**	Wiederverarbeitung *f*
1281	**reproducibility**	Reproduzierbarkeit *f*
1282	**residence time**	Verweilzeit *f*
1283	**residual elongation**	bleibende Dehnung *f*
1284	**residual melt cushion**	Restmassepolster *n*
1285	**residual monomer**	Restmonomer *n*
1286	**residual strength**	Restfestigkeit *f*
1287	**residual stroke**	Resthub *m*
1288	**resilience**	Rückstellvermögen *n*
1289	**resin**	Harz *n*; Kunstharz *n*
1290	**resin content**	Harzgehalt *m*
1291	**resin impregnated**	harzimprägniert
1292	**resin injection processing**	Harzinjektionsverfahren *n*
1293	**resin matrix**	Harzmatrix *f*
1294	**resin transfer moulding**	Harz-Transferpressen *n*

Français	*Español*	*Italiano*
retransformation *f*	reprocesado *m*	rilavorazione *f*
reproductibilité *f*	reproducibilidad *f*	riproducibilità *f*
temps *m* de séjour	tiempo *m* de permanencia	tempo *m* di sosta
élongation *f* résiduelle	alargamiento *m* residual	allungamento *m* residuo
coussin *m* de matière résiduel	cojín *m* de masa residual	massa *f* residua
monomère *m* résiduel	monómero *m* residual	monomero *m* residuo
contrainte *f* résiduelle	resistencia *f* residual	resistenza *f* residua
course *f* résiduelle	recorrido *m* residual	corsa *f* residua
résilience *f*	resiliencia *f*	resilienza *f*
résine *f*	resina *f*	resina *f*
taux *m* de résine	contenido *m* de resina	contenuto *m* di resina
imprégné de résine	impregnado de resina	impregnato con resina
injection *f* de résine	proceso *m* de inyección de resina	lavorazione *f* a iniezione di resina
résine *f* matrice	matriz *f* de resina	matrice *f* di resina
moulage *m* par transfert de résine	moldeo *m* por transferencia de resina	stampaggio *m* a trasferimento di resina

Nr.	English	Deutsch
1295	**resins for laminates**	Imprägnierharze *npl*
1296	**resistance**	Beständigkeit *f*
1297	**resol**	Resol *n*
1298	**resorcinol resin**	Resorcinharz *n*
1299	**restrictor bar**	Staubalken *m*
1300	**retardation**	Retardation *f*
1301	**retarder**	Verzögerer *m*
1302	**retreading**	Runderneuerung *f*
1303	**retrofit, to**	nachrüsten
1304	**reverse roll coating**	Umkehrwalzenbeschichtung *f*
1305	**rheogoniometer**	Rheogoniometer *n*
1306	**rheology**	Rheologie *f*
1307	**rheometry**	Rheometrie *f*
1308	**rheopexy**	Rheopexie *f*
1309	**rigid**	hart
1310	**rigid foam**	Hartschaumstoff *m*

Français	*Español*	*Italiano*
résines *fpl* pour lamination	resinas *fpl* de impregnación	resine *fpl* per stratificati
résistance *f*	resistencia *f;* estabilidad *f*	resistenza *f*
resol *m*	resol *m*	resol *m*
résine *f* résorcinique	resina *f* de resorcina	resina *f* resorcinia
barre *f* de régulation	barra *f* restrictora	barra *f* di arresto
retard *m*	retardación *f*	ritardo *m*
retardateur *m*	retardador *m*	agente *m* ritardante
rechapage *m*	recauchutado *m*	ricostruzione *f* (di pneumatici)
réhabiliter; rétrofitter	reconvertir	aggiornamento *m* (tecnicofunzionale)
enduction *f* par cylindre inverse	recubrimiento *m* por rodillos a contragiro	rivestimento *m* con rulli inversi
rhéogoniomètre *m*	reogoniómetro *m*	reogoniometro *m*
rhéologie *f*	reología *f*	reologia *f*
rhéométrie *f*	reometría *f*	reometrica *f*
rhéopexie *f*	reopexia *f*	reopessia *f*
rigide	rígido	rigido
mousse *f* rigide	espuma *f* rígida	espanso *m* rigido

Nr.	English	Deutsch
1311	**rigid PS foam**	Polystyrol-Hartschaum *m*
1312	**ring gate**	Ringanguss *m*
1313	**ring opening polymerisation**	ringöffnende Polymerisation *f*
1314	**riveting**	Nieten *n*
1315	**robot**	Roboter *m*
1316	**Rockwell hardness**	Rockwell-Härte *f*
1317	**rod**	Stab *m;* Stange *f*
1318	**roll**	Walze *f*
1319	**roll cover**	Walzenbelag *m*
1320	**roll mill**	Walzwerk *n;* Walzenstuhl *m*
1321	**roll-on machine**	Abrollmaschine *f*
1322	**roller**	Rolle *f;* Walze *f*
1323	**roller coating**	Walzenauftrag *m*
1324	**root surface** *(of a screw)*	Ganggrund *m*
1325	**rotary knife**	Kreismesser *n*

Français	Español	Italiano
mousse *f* rigide de PS	espuma *f* rígida de poliestireno	espanso *m* polistirenico rigido
injection *f* annulaire	entrada *f* anular	iniezione *f* anulare
polymérisation *f* par ouverture de cycle	polimerización *f* por apertura de anillo	polimerizzazione *f* con apertura ad anello
rivetage *m*	remache *m*	rivettatura *f*
robot *m*	robot *m*	robot *m*
dureté *f* Rockwell	dureza *f* Rockwell	durezza *f* Rockwell
baguette *f*	barra *f*	barra *f*
cylindre *m*	rodillo *m;* cilindro *m*	cilindro *m;* cilindratura *f*
revêtement *m* de cylindre	recubrimiento *m* del cilindro	rivestimento *m* del cilindro
broyeur *m* à cylindres	cilindros *mpl* mezcladores	mescolatore *m* a cilindri
machine *f* de déroulement	máquina *f* de desbobinado	avvolgitrice *f*
cylindre *m*	rodillo *m;* cilindro *m*	rullo *m*
enduction *f* à rouleaux	recubrimiento *m* por rodillo	rivestimento *m* a cilindro
surface *f* des filets de la vis	base *f* del filete	superficie *f* del solco
couteau *m* rotatif	cuchilla *f* giratoria	coltello *m* rotante

Nr.	English	Deutsch
1326	**rotary table**	Drehtisch *m*
1327	**rotary table machine**	Rundläufer *m*
1328	**rotate, to**	rotieren
1329	**rotational moulding; rotomoulding**	Rotationsformen *n*
1330	**rotational viscometer**	Rotationsviskosimeter *n*
1331	**roving**	Glasfaserstrang *m;* Roving *m*
1332	**rub fastness**	Reibechtheit *f*
1333	**rubber**	Kautschuk *m;* Gummi *m*
1334	**rubber elasticity**	Kautschuk-Elastizität *f*
1335	**rubberised fabric**	gummiertes Gewebe *n*
1336	**runner**	Angusskanal *m*
1337	**runner system**	Angusssystem *n*
1338	**runnerless**	angusslos
1339	**rupture disc**	Berstscheibe *f*
1340	**safety device**	Schutzvorrichtung *f*

Français	*Español*	*Italiano*
table *f* rotative	mesa *f* rotativa	tavola *f* rotante
table *f* rotative	máquina *f* de mesa rotativa	macchina *f* a giostra
tourner	girar	ruotare
rotomoulage *m*	moldeo *m* rotacional	stampaggio *m* rotazionale
viscosimètre *m* rotatif	viscosímetro *m* rotativo	viscosimetro *m* rotazionale
roving *m*	roving *m*	roving *m;* stoppino *m*
solidité *f* de friction	solidez *f* al frotamiento	resistenza *f* allo sfregamento
caoutchouc *m*	caucho *m;* goma *f*	gomma *f*
élasticité *f* du caoutchouc	elasticidad *f* del caucho	elasticità *f* della gomma
tissu *m* élastomérique	tejido *m* engomado	tessuto *m* gommato
canal *m* d'alimentation	canal *m* de alimentación	canale *m* principale
système *m* de canaux d'injection	sistema *m* de canales	sistema *m* di canali di colata
sans carotte *f*	sin bebedero *m*	senza canale *m* di colata
disque *m* de rupture	disco *m* de ruptura	disco *m* di rottura
dispositif *m* de sécurité	mecanismo *m* de seguridad	dispositivo *m* di sicurezza

Nr.	English	Deutsch
1341	**safety gate**	Schutzgitter *n*
1342	**sag**	Durchhängen *n*
1343	**sample**	Probe *f*
1344	**sampling**	Probenahme *f*
1345	**sandblasting**	Sandstrahlen *n*
1346	**sandwich**	Sandwich *m*
1347	**saponification number**	Verseifungszahl *f*
1348	**saturated**	gesättigt
1349	**saturation**	Sättigung *f*
1350	**scorching**	Anvulkanisieren *n*
1351	**scrap**	Abfall *m;* Ausschuss *m*
1352	**scratch hardness**	Ritzhärte *f*
1353	**scratch resistance**	Kratzfestigkeit *f*
1354	**screen**	Filtersieb *n*
1355	**screen changer**	Siebwechselvorrichtung *f*
1356	**screen pack**	Siebpaket *n*

Français	*Español*	*Italiano*
seuil *m* de sécurité	puerta *f* de seguridad	cancello *m* di sicurezza
fléchissement *m*	descolgamiento *m*	insaccamento *m*
échantillon *m*	muestra *f*	campione *m*
échantillonnage *m*	toma *f* de muestras	campionatura *f*
sablage *m*	chorro *m* de arena	sabbiatura *f*
sandwich *m*	sandwich *m*	sandwich *m*
nombre *m* de saponification	grado *m* de saponificación	valore *m* di saponificazione
saturé	saturado	saturato
saturation *f*	saturación *f*	saturazione *f*
brûlure *f*	vulcanizado *m*	scottatura *f*
déchet *m*	desperdicio *m*	scarto *m*
dureté *f* à la rayure	dureza *f* al rayado	durezza *f* della graffiatura
résistance *f* à la rayure	resistencia *f* al rayado	resistenza *f* al graffio
filtre *m*	filtro *m*	filtro *m*
changeur *m* de filtre	mecanismo *m* de cambio de filtros	cambiafiltro *m*
cartouche *f* filtrante	paquete *m* de filtros	pacco *m* filtro

Nr.	English	Deutsch
1357	screw	Schnecke *f*
1358	screw bushing	Schneckenbuchse *f*
1359	screw channel	Schneckengang *m;* Schneckenkanal *m*
1360	screw characteristic	Schneckenkennlinie *f*
1361	screw conveyor	Schneckenförderer *m*
1362	screw design	Schneckengeometrie *f*
1363	screw diameter	Schneckendurchmesser *m*
1364	screw drive	Schneckenantrieb *m*
1365	screw flight hard-facing	Schneckenpanzerung *f*
1366	screw joint	Schraubverbindung *f*
1367	screw length	Schneckenlänge *f*
1368	screw pitch	Gangsteigung *f*
1369	screw plastication	Schneckenplastifizierung *f*
1370	screw profile	Schneckenprofil *n*

Français	*Español*	*Italiano*
vis *f*	husillo *m*	vite *f*
bague *f* de vis	camisa *f* del husillo	boccola *f* della vite
canal *m* de vis	canal *m* del husillo	canale *m* della vite
caractéristiques *fpl* de la vis	característica *f* del husillo	caratteristica *f* della vite
dispositif *f* d'alimentation à vis	transportador *m* de husillo	convogliatore *m* della vite
conception *f* de la vis	geometría *f* del husillo	disegno *m* della vite
diamètre *m* de la vis	diámetro *m* del husillo	diametro *m* della vite
entraînement *m* de la vis	accionamiento *m* del husillo	azionamento *m* della vite
blindage *m* de la vis	templado *m* del husillo	filetto *m* con riporto duro
joint *m* de vis	unión *f* roscada	giunto *m* filettato
longueur *f* de la vis	longitud *f* del husillo	lunghezza *f* della vite
butée *f* de la vis	paso *m* del husillo	passo *m* della vite
plastification *f* par vis	plastificación *f* mediante husillo	plastificazione *f* a vite
profil *m* de la vis	perfil *m* del husillo	profilo *m* della vite

Nr.	English	Deutsch
1371	**screw retraction**	Schneckenrücklauf *m*
1372	**screw section**	Schneckenzone *f*
1373	**screw speed**	Schneckendrehzahl *f*
1374	**screw stroke**	Schneckenhub *m*
1375	**screw thread**	Schneckengewinde *n*
1376	**screw tip**	Schneckenspitze *f*
1377	**sealing**	Siegeln *n*
1378	**sealing area**	Siegelfläche *f*
1379	**sealing force**	Siegelkraft *f*
1380	**sealing layer**	Siegelschicht *f*
1381	**section drawing**	Schnittzeichnung *f*
1382	**segregation**	Absonderung *f*
1383	**self-extinguishing**	selbstverlöschend
1384	**self-ignition temperature**	Selbstentzündungs-temperatur *f*
1385	**self-reinforcement**	Eigenverstärkung *f*

Français	*Español*	*Italiano*
recul *m* de la vis	recorrido *m* de descompresión del husillo	arretramento *m* della vite
section *f* de la vis	zona *f* del husillo	sezione *f* della vite
vitesse *f* de vis	revoluciones *fpl* del husillo	giri *mpl* della vite
course *f* de la vis	carrera *f* del husillo	corsa *f* della vite
filet *m* de vis	filete *m* del husillo	filetto *m* della vite
sommet *m* de la vis	punta *f* del husillo	puntale *m* della vite
scellage *m*	sellado *m*	saldatura *f*
surface *f* de scellage	superficie *f* de sellado	superficie *f* di saldatura
force *f* de scellage	fuerza *f* de sellado	forza *f* di saldatura
couche *f* de scellage	capa *f* de sellado	strato *m* saldante
profil *m* de coupe	plano *m* de sección	disegno *m* di sezione
ségrégation *f*	segregación *f*	segregazione *f*
autoextinguible	autoextinguible	autoestinguente
température *f* d'auto inflammation	temperatura *f* de encendido espontáneo	temperatura *f* di autoignizione
auto renforcement *m*	autorrefuerzo *m*	autorinforzo *m*

Nr.	English	Deutsch
1386	**self-reinforcing polymer**	selbstverstärkender Kunststoff *m*
1387	**self-tapping screw**	gewindeschneidende Schraube *f*
1388	**semi-crystalline**	teilkristallin
1389	**semi-finished product**	Halbzeug *n*
1390	**semi-rigid**	halbhart
1391	**sensor**	Sensor *m;* Fühler *m*
1392	**service temperature**	Gebrauchstemperatur *f*
1393	**servo valve**	Servo-Ventil *n*
1394	**set-up** *(of a machine)*	Einrichten *n*
1395	**setting** *(of a machine)*	Einrichten *n*
1396	**setting time**	Abbindezeit *f*
1397	**setting-up time**	Rüstzeit *f*
1398	**shear**	Scherung *f*
1399	**shear deformation**	Scherdeformation *f*
1400	**shear flow**	Scherströmung *f*

Français	*Español*	*Italiano*
polymère *m* autorenforçant	polímero *m* autorreforzante	polimero *m* autorinforzante
vis *f* auto taraudeuse	tornillo *m* autocortante	vite *f* autofilettante
semi-cristallin	semicristalino	semicristallino
demi-produit *m*	producto *m* semiacabado	semilavorato *m*
semi-rigide	semirrígido	semirigido
capteur *m*	sensor *m*	sensore *m*
température *f* d'utilisation	temperatura *f* de servicio	temperatura *f* di esercizio
servovalve *f*	servoválvula *f*	servovalvola *f*
réglage *m*	instalación *f*	messa *f* a punto (di una macchina)
réglage *m*	ajustado *m*	messa *f* a punto
temps *m* de cuisson	tiempo *m* de curado	tempo *m* di reticolazione
temps *m* de réglage	tiempo *m* de puesta a punto	tempo *m* di messa a punto
cisaillement *m*	cizalla *f*	taglio *m*
déformation *f* par cisaillement	deformación *f* por cizalla	deformazione *f* di taglio
flux *f* au cisaillement	flujo *m* cizallante	flusso *m* di taglio

Nr.	English	Deutsch
1401	shear heating	Schererwärmung *f*
1402	shear modulus	Schubmodul *m*
1403	shear rate	Schergeschwindigkeit *f*
1404	shear strength	Scherfestigkeit *f*
1405	shear stress	Schubspannung *f*
1406	shear stretching	Scherverstreckung *f*
1407	shear velocity	Schergeschwindigkeit *f*
1408	shear viscosity	Scherviskosität *f*
1409	shear, to	abscheren
1410	sheet moulding compounds *(SMC)*	Harzmatten *fpl*
1411	sheeting; sheet	Platte *f*
1412	shelf life	Lagerbeständigkeit *f*
1413	shielding effect	Abschirmeffekt *m*

Français	Español	Italiano
chaleur *f* de cisaillement	calentamiento *m* por cizalla	riscaldamento *m* da taglio
module *m* de cisaillement	módulo *m* de cizalla	modulo *m* di taglio
taux *m* de cisaillement	velocidad *f* de cizalla	velocità di taglio
résistance *f* au cisaillement	resistencia *f* a la cizalla	resistenza *f* al taglio
contrainte *f* de cisaillement	esfuerzo *m* de cizalla	sollecitazione *f* al taglio
étirage *m* au cisaillement	estirado *m* de cizalla	estensione *f* al taglio
vitesse *f* de cisaillement	velocidad *f* de cizalla	velocità *f* di taglio
viscosité *f* de cisaillement	viscosidad *f* de cizalla	viscosità *f* di taglio
cisailler	cizallar	tagliare
sheet moulding compounds *mpl*	compuestos *mpl* de moldeo en lámina	composti *mpl* da stampaggio in lastra (SMC)
feuille *f*	placa *f*	lastra *f;* foglia *f*
durée *f* de vie au stockage	tiempo *m* máximo de almacenamiento	durata *f* a magazzino
effet *m* d'écran (ou de protection)	efecto *m* de apantallamiento	effetto *m* schermante

Nr.	English	Deutsch
1414	**Shore hardness**	Shore-Härte *f*
1415	**short-stroke press**	Kurzhubpresse *f*
1416	**short-term test**	Kurzzeitprüfung *f*
1417	**shot capacity**	Schussleistung *f*
1418	**shot volume**	Spritzvolumen *n;* Schussvolumen *n*
1419	**shot weight**	Schussgewicht *n*
1420	**shredder**	Abfallmühle *f*
1421	**shrink film**	Schrumpffolie *f*
1422	**shrink wrapping**	Schrumpffolienverpackung *f*
1423	**shrinkage**	Schwindung *f*
1424	**shut-off device**	Verschließeinrichtung *f*
1425	**shut-off nozzle**	Verschlussdüse *f*
1426	**shut-off valve**	Verschlussventil *n*
1427	**side chain**	Seitenkette *f*

Français	Español	Italiano
dureté *f* Shore	dureza *f* Shore	durezza *f* Shore
presse *f* à petite course	prensa *f* de recorrido corto	pressa *f* a corsa corta
essai *m* à court terme	ensayo *m* de corta duración	prova *f* a breve termine
capacité *f* d'injection	capacidad *f* de inyección	grammatura *f*
volume *m* injectable	volumen *m* de inyección	volume *m* d'iniezione
poids *m* injecté	peso *m* por embolada	peso *m* della stampata
déchiqueteur *m*	molino *m* de desperdicios	mulino *m* macinatore
film *m* rétractable	hoja *f* retráctil; lámina *f* retráctil	film *m* retraibile
emballage *m* rétractable	envase *m* de película retráctil	avvolgimento *m* con film retraibile
retrait *m*	contracción *f*	ritiro *m*
dispositif *m* de fermeture	dispositivo *m* de cierre	dispositivo *m* a serranda
buse *f* à obturation	boquilla *f* de válvula	ugello *m* valvola
clapet *m* (ou valve) de fermeture	válvula de cierre	valvola *f* a serranda
chaîne *f* latérale	cadena *f* lateral	catena *f* laterale

Nr.	English	Deutsch
1428	**side gate**	Seitenanguss *m*
1429	**sieve analysis**	Siebanalyse *f*
1430	**silane**	Silan *n*
1431	**silane-cured**	silanvernetzt
1432	**silica**	Kieselsäure *f*
1433	**silicone** *(SI)*	Silikon *n*
1434	**silicone resin**	Silikonharz *n*
1435	**silicone resin moulding compounds**	Silikonharz-Formmassen *fpl*
1436	**silicone rubber**	Silikonkautschuk *m*
1437	**silk screen printing**	Siebdruck *m*
1438	**silo**	Silo *m*
1439	**silvery streaks**	Silberschlieren *fpl*
1440	**simulation**	Simulation *f*
1441	**simulation program**	Simulationsprogramm *n*
1442	**single-cavity mould**	Einfachwerkzeug *n*

Français	Español	Italiano
injection *f* latérale	entrada *f* lateral	entrata *f* laterale
analyse *f* au crible	análisis *m* granulométrico	analisi *f* al setaccio
silane *m*	silano *m*	silano *m*
réticulé *f* au silane	reticulado *m* de silano	reticolazione *f* con silano
silice *f*	sílice *f*	silice *m*
silicone *m*	silicona *f*	silicone *m*
résine *f* silicone	resina *f* de silicona	resina *f* siliconica
matières *fpl* à mouler silicone	resinas *fpl* de silicona para moldeo	mescole *fpl* siliconiche da stampaggio
elastomère *f* de silicone; caoutchouc *f* silicone	caucho *m* de silicona	gomma *f* siliconica
sérigraphie *f*	serigrafía *f*	serigrafia *f*
silo *m*	silo *m*	silo *m*
traces *fpl* argentées	vetas *fpl* plateadas	segni *mpl* argentati
simulation *f*	simulación *f*	simulazione *f*
programme *m* de simulation	programa *f* de simulación	programma *m* di simulazione
moule *m* mono-empreinte	molde *m* de una cavidad	stampo *m* a una impronta

Nr.	English	Deutsch
1443	**single-flighted** *(screw)*	eingängig
1444	**single-layer film**	Einschichtfolie *f*
1445	**single screw**	Einfachschnecke *f*
1446	**single-screw extruder**	Einschneckenextruder *m*
1447	**sink marks**	Einfallstellen *fpl*
1448	**sintering**	Sintern *n*
1449	**size**	Schlichte *f*
1450	**skin pack**	Skinverpackung *f*
1451	**slabstock foam**	Blockschaumstoff *m*
1452	**sleeve ejector**	Hülsenauswerfer *m*
1453	**slide bar**	Schieber *m*
1454	**sliding bolt**	Schiebebolzen *m*
1455	**sliding friction**	Gleitreibung *f*
1456	**sliding guides**	Gleitführungen *fpl*
1457	**sliding shut-off nozzle**	Schiebeverschlussdüse *f*

Français	*Español*	*Italiano*
à un seul filet (vis)	de un sólo filete (tornillo)	con un solo filetto (vite)
film *m* monocouche	hoja *f* unicapa	film *m* monostrato
monovis *f*	monotornillo *m*	monovite *f*
extrudeuse *f* monovis	extrusora *m* de un tornillo	estrusore *m* monovite
retassures *fpl*	rechupados *mpl*	segni *mpl* di ritiro
frittage *m*	sinterizado *m*	sinterizzazione *f*
ensimage *f*	ensimaje *m*	bozzima *f*
skin pack *m*	envase *m* skin	imballaggio *m* a pelle
bloc *m* de mousse	bloque *m* de espuma	espanso *m* in blocchi
éjecteur *m* à manchon	casquillo *m* del expulsor	espulsore *m* tubolare
coulisseau *m*	barra deslizante *f*	barra *f* di scorrimento
coulisseau *m*	bulón *m* corredero	bullone *m* slittante
frottement *m* dynamique	fricción *f* deslizante	frizione *f* di scorrimento
guidages *mpl*	guías *fpl* deslizantes	guide *fpl* a slitta
obturateur *m* à coulisse	boquilla *f* de cierre hidráulica	ugello *m* a serranda scorrevole

Nr.	English	Deutsch
1458	**sliding split mould**	Schieberwerkzeug *n*
1459	**sliding table**	Schiebetisch *m*
1460	**slit die; slot die**	Breitschlitzdüse *f*
1461	**slitting machine**	Spaltmaschine *f*
1462	**slurry process**	Aufschlämmungsprozess *m*
1463	**slush mould** *(plastisols)*	Gießwerkzeug *n*
1464	**smectic**	smektisch
1465	**snap fit**	Schnappverbindung *f*
1466	**socket fusion welding**	Muffenschweißen *n*
1467	**softening point**	Erweichungspunkt *m*
1468	**sol**	Sol *n*
1469	**solder bath**	Lötbad *n*
1470	**solid**	fest
1471	**solid flow**	Blockströmung *f*
1472	**solubility**	Löslichkeit *f*

Français	*Español*	*Italiano*
moule *m* à tiroir	molde *m* de corredera	stampo *m* a slitta
table *f* coulissante	mesa *f* desplazable	tavola *f* scorrevole
filière *f* plate	boquilla *f* plana	filiera *f* piana; testa *f* piana
machine *f* de découpe	máquina *f* de cortar	tranciatrice *f*
procédé *m* en suspension	proceso *m* slurry	processo *m* in soluzione
moule *m* pour plastisols	molde *m* de embarrado	stampo *f* per plastisol
smectique	smectico	smettico
joint *m* instantané	unión *f* por presión	assemblaggio *m* a scatto
soudure *f* au cordon chauffant	soldadura *f* por manguitos	saldatura *f* per fusione a incastro
température *f* de ramollissement	temperatura *f* de reblandecimiento	temperatura *f* di rammollimento
colloïdal *m*	solución *f* coloidal	sol *m*
bain *m* de soudure	baño *m* de soldar	bagno *m* di saldatura
solide	sólido	compatto
gaine *f* solide	flujo *m* sólido	scorrimento *m* a solido
solubilité *f*	solubilidad *f*	solubilità *f*

Nr.	English	Deutsch
1473	**soluble core technique**	Lösekern-Technologie *f*
1474	**solution**	Lösung *f*
1475	**solution polymerisation**	Lösungspolymerisation *f*
1476	**solution properties**	Lösungseigenschaften *fpl*
1477	**solvent**	Lösungsmittel *n*
1478	**solvent resistance**	Lösemittelbeständigkeit *f*
1479	**solvent welding**	Quellschweißen *n*
1480	**sorption**	Sorption *f*
1481	**sound absorption**	Schalldämmung *f*
1482	**sound insulation**	Schallisolierung *f*
1483	**sound protection**	Schallschutz *m*
1484	**sound reduction**	Schalldämpfung *f*
1485	**spacer plates**	Distanzplatten *fpl*
1486	**spacer rod**	Distanzstange *f*

Français	*Español*	*Italiano*
technique *f* aux noyaux solubles	técnica *f* de núcleos solubles	tecnica *f* con nucleo a perdere
solution *f*	solución *f*	soluzione *f*
polymérisation *f* en solution	polimerización *f* en solución	polimerizzazione *f* in soluzione
propriétés *fpl* en solution	propiedades *mpl* de solución	proprietà *fpl* in soluzione
solvant *m*	disolvente *m*	solvente *m*
résistance *f* au solvant	resistencia *f* a los disolventes	resistenza *f* ai solventi
soudage *f* au solvant	soldadura *f* con disolventes	saldatura *f* con solvente
adsorption *f*	adsorción *f*	sorbimento *m*
absorption *f* du son	absorción *f* acústica	assorbimento *m* del suono
isolation *f* phonique	aislamiento *m* acústico	isolamento *m* acustico
protection *f* antibruit	protección *f* acústica	protezione *f* dal rumore
atténuation *f* du son	amortiguamiento *m* acústico	riduzione *f* del rumore
plaques *fpl* de séparation	placas *fpl* distanciadoras	distanziatori *mpl*
bague *f* de séparation	varilla *f* de separación	asta *f* distanziale

Nr.	English	Deutsch
1487	**spark erosion**	Funkenerosion *f*
1488	**specific gravity**	spezifisches Gewicht *n*
1489	**specific heat**	spezifische Wärme *f*
1490	**specific viscosity**	spezifische Viskosität *f*
1491	**specific volume**	spezifisches Volumen *n*
1492	**specification**	Spezifikation *f*
1493	**specimen**	Probekörper *m*
1494	**speck**	Stippe *f*
1495	**spectroscopy**	Spektroskopie *f*
1496	**speed control**	Drehzahlregelung *f*
1497	**spherical molecule**	Kugelmolekül *n*
1498	**spherulite**	Sphärolith *m*
1499	**spider**	Dornsteghalter *m*
1500	**spider lines**	Dornhaltermarkierungen *fpl*
1501	**spin welding**	Rotationsreibschweißen *n*

Français	Español	Italiano
électroérosion *f*	electroerosión *f*	elettroerosione *f*
masse *f* volumique	peso *m* específico	peso *m* specifico
chaleur *f* spécifique	calor *m* específico	temperatura *f* specifica
viscosité *f* intrinsèque	viscosidad *f* específica	viscosità *f* specifica
volume *m* spécifique	volumen *m* específico	volume *m* specifico
spécification *f*	especificación *f*	specifica *f*
spécimen *m*	probeta *f*	campione *m*
tache *f*	partícula *f*	macchia *f*
spectroscopie *f*	espectroscopia *f*	spettroscopia *f*
régulation *f* de la vitesse	regulador *m* de revoluciones	controllo *m* di velocità
molécule *f* sphérique	molécula *f* esférica	molecola *f* sferica
sphérulite *f*	esferolita *f*	sferulite *f*
porte-poinçon *m*	portamandril *m*	crociera *f*
marques *fpl* de mandrin	marcas *fpl* producidas por el portamandril	segni *mpl* di crociera
soudage *m* par friction	soldeo *m* por fricción rotacional	saldatura *f* a frizione

Nr.	English	Deutsch
1502	**spinneret**	Spinndüse *f*
1503	**spiral conveyor**	Spiralförderer *m*
1504	**spiral flow length**	Spirallänge *f*
1505	**spiral mandrel** *(melt)* **distributor**	Wendelverteiler *m*
1506	**split mould**	Backenwerkzeug *n*
1507	**spot welding**	Punktschweißen *n*
1508	**spray cooling**	Sprühkühlung *f*
1509	**spray lay-up**	Faserspritzen *n*
1510	**spray lay-up moulding**	Faserspritzverfahren *n*
1511	**spraygun**	Spritzpistole *f*
1512	**spread coating**	Streichverfahren *n*
1513	**spreader pin**	Spreizdorn *m*
1514	**spreader roll unit**	Breitstreckwerk *n*

Français	Español	Italiano
filière *f* de filage	hilera *f* de filamentos	filiera *f*
transporteuse *f* à spirale	transportador *m* de espiral	convogliatore *m* a spirale
longueur *f* du flux en spirale	longitud *f* de la espiral	lunghezza *f* della spirale
distributeur *m* de mandrin spiralé	distribuidor *m* de inversión (mandril de espiral)	distributore *m* con mandrino a spirale
moule *m* à coins	molde *m* partido	stampo *m* diviso
soudure *f* par points	soldadura *f* por puntos	saldatura *f* a punti
refroidissement *m* par projection	enfriamiento *m* por rociado	raffreddamento *m* a spruzzo
projection *f* de fibres	proyección *f* de fibras	spalmatura *f* a spruzzo
moulage *m* par projection de fibres	moldeo *m* por proyección de fibra	iniezione *f* di fibra
pistolet *m* de projection	pistola *f* de proyección	pistola *f* spruzzatrice
revêtement *m* par nappage	recubrimiento *m* por extensión	rivestimento *m* per spalmatura
mandrin *m* cylindrique	punzón *m* de expansión	perno *m* per spalmatura
cylindre *m* d'étirement	esparcidora *f* de rodillos	rullo *m* spalmatore

Nr.	English	Deutsch
1515	**sprue**	Anguss *m*
1516	**sprue bush**	Angussbuchse *f*
1517	**sprue gate**	Kegelanguss *m*
1518	**sprueless**	angusslos
1519	**sprueless injection moulding**	angussloses Spritzgießen *n*
1520	**stabiliser**	Stabilisator *m*
1521	**stack mould**	Etagenwerkzeug *n*
1522	**stacking**	Stapeln *n*
1523	**stacking device**	Stapelvorrichtung *f*
1524	**stamping**	Prägen *n*
1525	**stamping foil**	Prägefolie *f*
1526	**stamping press**	Prägepresse *f*
1527	**standard elements** *(mould)*	Werkzeug-Normalien *fpl;* Normalien *fpl*

Français	*Español*	*Italiano*
carotte *f*	bebedero *m*	materozza *f;* canale *m* principale di iniezione
douille *f* de carotte	manguito *m* del bebedero	boccola *f* di iniezione
carotte *f* en cône	mazarote *m* cónico	materozza *f* a cono
sans carotte	sin bebedero	senza materozza
injection *f* sans carotte	inyección *f* sin bebedero	stampaggio *m* a iniezione senza materozza
stabilisant *m*	estabilizador *m*	stabilizzante *m*
moule *m* à étages	molde *m* de pisos; molde *m* de libro	stampo *m* a piani
empilement *m*	apilamiento *m*	impilamento *m*
empileuse *f*	mecanismo *m* de apilamiento	impilatrice *f*
poinçonnage *m*; estampage *m*	estampado *m*	stampatura *f*
film *m* pour impression à chaud	cinta *f* para estampado	foglia *f* per impressione a caldo
machine *f* de marquage à chaud	prensa *f* de estampar	pressa *f* per impressione a caldo
éléments *mpl*, standard de moule	unidades *fpl* de molde normalizadas	elementi *mpl* normalizzati per stampi

Nr.	English	Deutsch
1528	**standard equipment**	Standardausführung *f*
1529	**standard mould unit**	Stammwerkzeug *n*
1530	**standard screw**	Standardschnecke *f*
1531	**standard test piece**	Normstab *m*
1532	**standards**	Normen *fpl*
1533	**staple fibre**	Stapelfaser *f*
1534	**starch copolymers**	Copolymere *npl* auf Stärkebasis
1535	**start, to**	anfahren
1536	**state of aggregation**	Aggregatzustand *m*
1537	**static friction**	Haftreibung *f*
1538	**static mixer**	Statikmischer *m*
1539	**stationary**	feststehend
1540	**stepless**	stufenlos
1541	**stereolithography**	Stereolithografie *f*
1542	**stick-slip effect**	Haft-Gleit-Effekt *m*

Français	*Español*	*Italiano*
équipement *m* standard	equipamiento *m* estándar	attrezzatura *f* standard
carcasse *f* de moule	molde *m* normalizado	stampo *m* base
vis *f* standard	husillo *m* estándar	vite *f* standard
éprouvette *f* normalisée	probeta *f* normalizada	campione *m* normalizzato
normes *fpl*	normas *fpl*	norme *fpl*
fibre *f* brute	fibra *f* corta	fibra *f* discontinua
copolymères *mpl* d'amidon	copolímeros *mpl* basados en almidón	copolimeri *mpl* di amido
faire démarrer	poner en marcha	avviare; mettere in funzione
état *m* d'aggrégation	estado *m* de agregación	stato *m* di aggregazione
frottement *m* statique	fricción *f* estática	frizione *f* statica
mélangeur *m* statique	mezclador *m* estático	mescolatore *m* statico
fixe	fijo; estacionario	fisso
sans étape	sin escalonado	senza limiti
stéréolithographie *f*	estereolitografía *f*	stereolitografia *f*
effet *m* collant/glissant	efecto *m* deslizamiento-adhesión	effetto *m* adesione-scivolamento

Nr.	English	Deutsch
1543	**sticking**	Anbacken *n*
1544	**stiffness**	Steifheit *f*
1545	**stop pin**	Anschlagbolzen *m*
1546	**storage modulus**	Speichermodul *m*
1547	**straight-through die** *(extr.)*	Längsspritzkopf *m*
1548	**strain**	Dehnung *f* (unter Spannung)
1549	**strainer**	Lochscheibe *f*
1550	**strand pelletiser**	Stranggranulator *m*
1551	**streaking**	Schlierenbildung *f*
1552	**streaks**	Schlieren *fpl*
1553	**strength**	Festigkeit *f*
1554	**stress**	Spannung *f*
1555	**stress analysis**	Spannungsanalyse *f*
1556	**stress birefringence**	Spannungsdoppelbrechung *f*

Français	*Español*	*Italiano*
collant *m*	pegado *m*	appiccicato *m*
rigidité *f*	rigidez *f*	rigidità *f*
obturateur *m* à aiguille	perno *m* de tope	perno *m* di arresto
module *m* de stockage	módulo *m* de almacenamiento	modulo *m* di permanenza della massa
filière *f* directe	boquilla *f* recta	filiera *f* diritta
allongement *m* (tension)	alargamiento *m* (esfuerzo)	allungamento *m* (su sollecitazione)
grille *f*	plato *m* rompedor	estrusore *m* filtrante
granulateur *m* à bande	granuladora *f* de hilos	granulatore *m*
rayure *f*	formación *f* de lineas	striatura *f*
stries *f*	líneas *fpl;* ráfagas *fpl* superficiales	striature *fpl*
résistance *f*	resistencia *f*	resistenza *f*
tension *f;* contrainte *f*	esfuerzo *m*	sollecitazione *f*
analyse *f* des contraintes	análisis *m* de esfuerzo	analisi *f* della tensione
biréfringence *f* sous contraintes	esfuerzo *m* birrefringente	birifrangenza *f* alla sollecitazione

Nr.	English	Deutsch
1557	**stress cracking**	Spannungsrissbildung *f*
1558	**stress-free**	spannungsfrei
1559	**stress relaxation**	Spannungsrelaxation *f*
1560	**stress-strain diagram**	Spannungs-Dehnungs-Diagramm *n*
1561	**stress whitening**	Weißbruch *m*
1562	**stretch blow moulding**	Streckblasformen *n*
1563	**stretch film**	Stretchfolie *f*
1564	**stretch wrapping**	Streckfolienverpackung *f*
1565	**stretch, to**	verstrecken
1566	**stretching**	Recken *n*
1567	**stretching roll**	Reckwalze *f*
1568	**stretching unit**	Reckanlage *f*
1569	**stripper**	Abstreifer *m*
1570	**stripper bar**	Abstreifleiste *f*

Français	*Español*	*Italiano*
fissuration *f* sous contrainte	agrietamiento *m* por esfuerzo	fessurazione *f* sotto sforzo
libre de contraintes	sin tensiones	senza sollecitazioni
relaxation *f* des contraintes	esfuerzo *m* de relajación	rilassamento *m* in sollecitazione
diagramme *m* contrainte allongement	diagrama *m* deformación esfuerzo	diagramma *m* di sollecitazione
blanchiment *m* à la contrainte	fractura *f* blanca	sbiancamento *m* da sollecitazione
soufflage *m* par biorientation	moldeo *m* por estirado-soplado	soffiaggio *m* stiro
film *m* étirable	hoja *f* estirable; lámina *f* estirable	film *m* estensibile
emballage *m* étirable	embalaje *m* por estiraje de hoja	avvolgimento *m* con film estensibile
étirer	estirar	stirare
étirage *m*	estirado *m;* orientación *f*	stiro *m*
rouleau *m* d'étirage	cilindro *m* de estiraje	rullo *m* di stiro
unité *f* d'étirage	instalación *f* de estiraje	unità *f* di stiro
extracteur *m*	extractor *m*	estrattore *m*
tige *f* d'éjection	vara *f* de extracción	barra *f* di sfilamento

Nr.	English	Deutsch
1571	**stripper plate**	Abstreifplatte *f*
1572	**stripper ring**	Abstreifring *m*
1573	**stroke**	Hub *m*
1574	**stroke limit**	Hubbegrenzung *f*
1575	**structural foam**	Integralschaumstoff *m;* Strukturschaumstoff *m*
1576	**structural foam moulding**	Strukturschaumgießen *n;* TSG-Verfahren *n*
1577	**stuffing unit; stuffer**	Stopfvorrichtung *f*
1578	**styrene**	Styrol *n*
1579	**styrene acrylonitrile copolymer** *(SAN)*	Styrol-Acrylinil-Copolymer *n*
1580	**styrene block copolymer**	Styrol-Blockcopolymer *n*
1581	**styrene butadiene copolymer** *(SB)*	Styrol-Butadien-Copolymer *n*
1582	**styrene copolymer**	Styrol-Copolymer *n*
1583	**styrene polymers**	Styrolpolymere *npl*
1584	**submarine gate**	Tunnelanschnitt *m*

Français	*Español*	*Italiano*
plaque *f* d'extraction	placa *f* de extracción	piastra *f* di estrazione
anneau *m* d'éjection	aro *m* de extracción	anello *m* di sfilamento
course *f*	recorrido *m;* carrera *f*	corsa *f*
butée *f*	límite *m* de la carrera	fine *f* corsa
mousse *f* structurale	espuma *f* integral	espanso *m* strutturale
pièces *fpl* en mousse structurale	moldeo *m* de espuma estructural	stampaggio *m* di espanso strutturale
bourreur *m*	dispositivo *m* de obturación	premistoppa *m*
styrène *m*	estireno *m*	stirene *m*
copolymère *m* de styrène-acrylonitrile	copolímero *m* de estireno-acrilonitrilo	copolimero *m* stirene acrilonitrile
copolymère *m* styrénique à blocs	copolímero *m* de bloque de estireno	copolimero *m* a blocchi stirenici
copolymère *m* de styrène butadiène	copolímero *m* de estireno-butadieno	copolimero *m* stirene butadiene
copolymère *m* styrénique	copolímero *m* de estireno	copolimero *m* stirenico
polymères *mpl* styréniques	polímeros *mpl* estirénicos	polimeri *mpl* stirenici
injection *f* sous-marine	entrada *f* submarina	punto *m* di iniezione sommerso

Nr.	English	Deutsch
1585	**suction conveyor**	Saugfördergerät *n*
1586	**suction feeder**	Saugförderer *m*
1587	**supercritical fluid**	superkritische Flüssigkeit *f*
1588	**support ring**	Stützring *m*
1589	**supporting film**	Trägerfolie *f*
1590	**surface coating**	Oberflächenbeschichtung *f*
1591	**surface finish**	Oberflächengüte *f*
1592	**surface gloss**	Oberflächenglanz *m*
1593	**surface layer**	Oberflächenschicht *f*
1594	**surface resistance**	Oberflächenwiderstand *m*
1595	**surface treatment**	Oberflächenbehandlung *f*
1596	**surfactants**	Tenside *npl*
1597	**suspension polymerisation**	Suspensionspolymerisation *f*

Français	*Español*	*Italiano*
aspiration *f* par succion	transporte *m* por succión	convogliatore *m* aspirante
dispositif *m* d'alimentation par succion	transporte *m* por succión	alimentatore *m* aspirante
fluide *m* supercritique	fluido *m* supercrítico	fluido *m* supercritico
anneau *m* de support	anillo *m* de sporte	anello *m* di supporto
feuille *f* de transport	hoja *f* soporte	film *m* di supporto
couchage *m*	recubrimiento *m* superficial	verniciatura *f;* laccatura *f* superficiale
fini *m* de surface	acabado *m* superficial	finitura *f* superficiale
brillant *m* de surface	brillo *m* superficial	brillantezza *f* superficiale
couche *f* superficielle	capa *f* superficial	strato *m* superficiale
résistance *f* de la surface	resistencia *f* superficial	resistenza *f* superficiale
traitement *m* de surface	tratamiento *m* superficial	trattamento *m* superficiale
surfactants *mpl*	tensioactivos *mpl;* surfactantes *mpl*	tensioattivi *mpl*
polymérisation *f* en suspension	polimerización *f* en suspensión	polimerizzazione *f* in sospensione

Nr.	English	Deutsch
1598	**swelling**	Quellung *f*
1599	**swelling index**	Quellungsindex *m*
1600	**swivel-out**	ausschwenkbar
1601	**swivel-type injection unit**	schwenkbares Spritzaggregat *n*
1602	**syndiotactic**	syndiotaktisch
1603	**syndiotactic polymers**	syndiotaktische Polymere *npl*
1604	**synthesis**	Synthese *f*
1605	**synthetic fibre**	Synthesefaser *f*
1606	**synthetic leather**	Kunstleder *n*
1607	**synthetic resin**	Kunstharz *n*
1608	**synthetic rubber**	Synthesekautschuk *m*
1609	**tack**	Klebrigkeit *f*
1610	**tackifier**	Klebrigmacher *m*
1611	**tacticity**	Taktizität *f*
1612	**take-off**	Abzug *m*
1613	**take-off rolls**	Abzugswalze *f*

Français	Español	Italiano
gonflement *m*	hinchamiento *m*	rigonfiamento *m*
taux *m* de gonflement	índice *m* de hinchamiento	indice *m* di rigonfiamento
pivotant	basculante	tavola rotante
unité *f* d'injection pivotante	unidad *f* de inyección basculante	unità *f* di iniezione rotante
syndiotactique	sindiotáctico	sindiotattico
polymères *mpl* syndiotactiques	polimeros *mpl* sindiotácticos	polimeri *mpl* sindiotattici
synthèse *f*	síntesis *f*	sintesi *f*
fibre *f* synthétique	fibra *f* sintética	fibra *f* sintetica
cuir *m* synthétique	cuero *m* artificial	pelle *m* artificiale
résine *f* synthétique	resina *f* sintética	resina *f* sintetica
caoutchouc *m* synthétique	caucho *m* sintético	gomma *f* sintetica
tack *m*; collabilité *f*	pegajosidad *f*	appiccicosità *f*
tackifiant *m*	agente *m* de pagajosidad	adesivante *m*
tacticité *f*	tacticidad *f*	tattica *f*
tirage *m*	estiraje *m*	traino *m*
cylindre *m* de tirage; dérouleur *m*	rodillo *m* del tren de estiraje	rulli *mpl* di traino

Nr.	English	Deutsch
1614	**take-off speed**	Abzugsgeschwindigkeit *f*
1615	**take-off tension**	Abzugskraft *f*
1616	**take-off unit**	Abzugsvorrichtung *f*
1617	**talc**	Talkum *n*
1618	**tapered**	kegelförmig
1619	**tear propagation resistance**	Weiterreißfestigkeit *f*
1620	**tear propagation test**	Weiterreißversuch *m*
1621	**tear resistance**	Einreißfestigkeit *f*
1622	**tear resistance test**	Einreißversuch *m*
1623	**tear strength**	Reißfestigkeit *f*
1624	**tempered steel**	vergüteter Stahl *m*
1625	**tensile strength**	Zugfestigkeit *f*
1626	**tensile stress**	Zugspannung *f*

Français	*Español*	*Italiano*
vitesse *f* de tirage	velocidad *f* de estiraje	velocità *f* di stiro
tension *f* de tirage	fuerza *f* de estiraje	tensione *f* di tirata
unité *f* de tirage	tren *m* de arrastre	unità *f* di tirata
talc *m*	talco *m*	talco *m*
cônique	cónico	rastremato
résistance *f* à la propagation de la déchirure	resistencia *f* a la propagación del desgarro	resistenza *f* alla propagazione dello strappo
test *m* de propagation de la déchirure	ensayo *m* de propagación al desgarro	prova *f* di propagazione della lacerazione
résistance *f* à la déchirure	resistencia *f* al desgarro	resistenza *f* allo strappo
test *m* de résistance au déchirement	ensayo *m* de resistencia al desgarro	prova *f* di resistenza alla lacerazione
résistance *f* au déchirement	resistencia *f* al desgarro	resistenza *f* alla lacerazione
acier *m* trempé	acero *m* mejorado	acciaio *m* bonificato
module *m* de traction	resistencia *f* a la tracción	resistenza *f* alla trazione
contrainte *f* en traction	esfuerzo *m* de tracción	sollecitazione *f* per trazione

Nr.	English	Deutsch
1627	**tensile test**	Zugversuch *m*
1628	**terpolymer**	Terpolymer *n*
1629	**test**	Prüfung *f*
1630	**test specimen**	Probekörper *m*
1631	**thermal conductivity**	Wärmeleitfähigkeit *f*
1632	**thermal degradation**	thermischer Abbau *m*
1633	**thermal expansion**	thermische Ausdehnung *f*
1634	**thermal sealing**	Wärmekontaktschweißen *n*
1635	**thermoanalysis; thermal analysis**	Thermoanalyse *f*
1636	**thermoelasticity**	Thermoelastizität *f*
1637	**thermoforming**	Warmformen *n;* Thermoformen *n*
1638	**thermoplastic elastomers**	thermoplastische Elastomere *npl*
1639	**thermoplastic resin**	Thermoplast *m*
1640	**thermoset**	Duroplast *m*
1641	**thermosetting**	hitzehärtbar

Français	*Español*	*Italiano*
test *m* de traction	ensayo *m* de tracción	prova *f* di trazione
terpolymère *m*	terpolímero *m*	terpolimero *m*
essai *m*	ensayo *m*	prova *f*
éprouvette *f*	probeta *f*	provetta *f*
conductivité *f* thermique	conductividad *f* térmica	conduttività *f* termica
dégradation *f* thermique	degradación *f* térmica	degradazione *f* termica
dilatation *f* thermique	dilatación *f* térmica	dilatazione *f* termica
soudage *m* au contact	soldadura *f* por impulso de calor	saldatura *f* a caldo
thermoanalyse *f*	análisis *m* térmico	termoanalisi *f*
thermoélasticité *f*	elasticidad *f* térmica	termoelasticità *f*
thermoformage *m*	termoconformado *m*	termoformatura *f*
élastomères *mpl* thermoplastiques	elastómeros *mpl* termoplásticos	elastomeri *mpl* termoplastici
thermoplastique *m*	termoplástico *m*	resina *f* termoplastica
thermodurcissable *m*	plástico *m* termoestable	termoindurente *m*
thermodurcissable	termoendurecible	termoindurente

Nr.	English	Deutsch
1642	**thermotropic**	thermotrop
1643	**thickening agent; thickener**	Verdickungsmittel *n*
1644	**thickness gauge**	Dickenmesseinrichtung *f*
1645	**thin-wall**	dünnwandig
1646	**thin-wall technology**	Dünnwandtechnik *f*
1647	**thixotropic**	thixotrop
1648	**thread**	Gewinde *n*
1649	**thread insert**	Gewindeeinsatz *m*
1650	**threaded spindle**	Gewindespindel *f*
1651	**three-plate mould**	Dreiplattenwerkzeug *n*
1652	**three-roll calender**	Dreiwalzenkalander *m*
1653	**three-zone screw**	Dreizonenschnecke *f*
1654	**throttle**	Drossel *f*
1655	**throttle valve**	Drosselventil *n*
1656	**throughput**	Durchsatz *m*

Français	*Español*	*Italiano*
thermotrope	termotrópico	termotropico
agent *m* épaississant	agente *m* espesante	agente *m* addensante
jauge *f* d'épaisseur	medidor *m* de espesor	spessore *m*
à paroi *f* fine	pared *f* delgada	parete *f* sottile
technologie *f* pour parois fines	técnica *f* de pared delgada	tecnologia *f* a parete sottile
thixotropique	tixotrópico	tissotropico
fil *m*	filete *m*	filo *m*
insert *m* fileté	inserto *m* roscado	inserto *m* filettato
tige *f* filetée	sinfín *m* roscado	mandrino *m* filettato
moule *m* à trois plateaux	molde *m* de tres placas	stampo *m* con piastra intermedia
calandre *f* à trois cylindres	calandra *f* de tres cilindros	calandra *f* a tre cilindri
vis *f* à trois zones	husillo *m* de tres zonas	vite *f* a tre zone
clapet *m*	estrangulador *m*	valvola *f* a farfalla
soupape *m* d'étranglement; papillon *m* de réglage	válvula *f* estranguladora	valvola *f* a strozzamento
débit *m;* production *f*	capacidad *f;* volumen *m* de descarga	capacitá *f*

Nr.	English	Deutsch
1657	**throughput rate**	Förderleistung *f*
1658	**thrust rod**	Schubstange *f*
1659	**tie-bar**	Holm *m*
1660	**tie-bar distance**	Holmabstand *m*
1661	**tie-barless**	holmlos
1662	**toggle**	Kniehebel *m*
1663	**toggle press**	Kniehebelpresse *f*
1664	**tooling costs**	Werkzeugkosten *fpl*
1665	**torpedo**	Torpedo *m*
1666	**torque of screw**	Schneckendrehmoment *n*
1667	**torsion pendulum test**	Torsionsschwingungs-versuch *m*
1668	**toughening agent**	Schlagzähigkeits-verbesserer *m*
1669	**toughness**	Zähigkeit *f*
1670	**tracking resistance**	Kriechstromfestigkeit *f*

Français	Español	Italiano
rendement *m*	índice *m* de rendimiento	capacità *f* massima
bielle *f*	varilla *f* de empuje	asta *f* di spinta
colonne *f*	columna *f;* larguero *m*	barra *f* di bloccaggio
distance *f* entre colonnes	distancia *f* entre columnas	distanza *f* fra le colonne
sans colonnes	sin columnas	senza colonne
genouillère *f*	palanca *f* acodada; rodillera *f*	ginocchiera *f*
presse *f* à genouillère	prensa *f* de palanca acodada	pressa *f* a ginocchiera
coûts *mpl* d'outillage	costes *mpl* del molde	costi *mpl* di attrezzaggio
torpille *f*	torpedo *m*	testa *f;* siluro *m*
couple *m* de la vis	par *m* del husillo	coppia *f* della vite
test *m* de torsion au pendule	ensayo *m* de torsión oscilante	prova *f* pendolo di torsione
modifiant *m* choc	modificador *m* de impacto	agente *m* tenacizzante
ténacité *f*	tenacidad *f*	tenacità *f*
résistance *f* au fluage	resistencia *f* al tracking	resistenza *f* alle correnti striscianti

Nr.	English	Deutsch
1671	**transfer mould**	Spritzpresswerkzeug *n*
1672	**transfer moulding**	Spritzpressen *n*
1673	**transfer moulding press**	Spritzpresse *f*
1674	**translucent**	transluzent
1675	**transparent**	transparent
1676	**transverse flow**	Querströmung *f*
1677	**tribo-chemical**	tribochemisch
1678	**tribo-rheological**	triborheologisch
1679	**trimming machine**	Entgratungsmaschine *f*
1680	**triple-flighted**	dreigängig
1681	**triple roller**	Dreiwalzenstuhl *m*
1682	**triple screw**	Dreifachschnecke *f*
1683	**tube**	Rohr *n*
1684	**tubular die**	Ringdüse *f*
1685	**tubular film**	Schlauchfolie *f*

Français	*Español*	*Italiano*
moule *m* transfert	molde *m* de transferencia	stampo *m* di trasferimento
moulage *m* par transfert	moldeo *m* por transferencia	stampaggio *m* a transfer
presse *f* transfert	prensa *f* de moldeo por transferencia	pressa *f* per stampaggio a transfer
translucide	translúcido	translucido
transparent	transparente	trasparente
flux *m* transverse	flujo *m* transversal	flusso *m* trasversale
tribochimique	triboquímico	tribochimico
tribo-rhéologique	triboreológico	triboreologico
machine *f* à ébavurer	máquina *f* de desbarbado	macchina *f* rifilatrice
à trois voies	de filete triple	triplofilettata
triple cylindre *m*	tricilíndrica *f*	triplo rullo *m*
triple vis *f*	tornillo triple *m*	vite *f* tripla
tuyau *m;* tube *m*	tubo *m*	tubo *m*
filière *f* annulaire	tobera *f* anular; hilera *f* anular	filiera *f* anulare
film *m* tubulaire	hoja *f* tubular	film *m* tubolare

Nr.	English	Deutsch
1686	**tumble dryer**	Trommeltrockner *m*
1687	**tumble mixer**	Taumelmischer *m*
1688	**tunnel gate**	Tunnelanguss *m*
1689	**turbidity**	Trübung *f*
1690	**twin-screw extruder**	Doppelschneckenextruder *m*
1691	**twin-sheet thermoforming**	Twin-Sheet-Thermoformen *n*
1692	**twin-wall sheet**	Doppelstegplatte *f*
1693	**two-colour injection moulding**	Zweifarben-Spritzgießen *n*
1694	**two-stage**	zweistufig
1695	**ultimate tensile strength**	Reißfestigkeit *f*
1696	**ultra-high pressure plasticising**	Höchstdruck-Plastifizierung *f*
1697	**ultrasonic welding**	Ultraschallschweißen *n*
1698	**undercut**	Hinterschneidung *f*

Français	*Español*	*Italiano*
sécheur *m* à tambour	secador *m* de tambor	essiccatore *m* a tamburo
mélangeur *m* secoueur	mezclador *m* rotativo basculante	mescolatore *m* a tamburo
injection *f* sous marine	entrada *f* submarina; entrada *f* de túnel	orifizio *m* a tunnel
turbidité *f*	turbidez *f*	torbidità *f*
extrudeuse *f* double vis	extrusora *m* de dos tornillos	estrusore *m* a doppia vite
thermoformage *m* en double feuille	termoconformado *m* de doble lámina	termoformatura *f* a doppia lastra
feuille *f* double	placa *f* con nervadura doble	lastra *f* a doppia parete
injection *f* bicolore	inyección *f* de dos colores	iniezione *f* a due colori
à deux étapes	de dos pasos; de dos escalones	a due stadi
traction *f* à la rupture	resistencia *f* a la tracción última	resistenza *f* limite alla trazione
plastification *f* à très haute pression	plastificación *f* mediante presión ultra alta	plastificazione *f* a pressione ultraelevata
soudage *m* par ultrasons	soldadura *f* por ultrasonidos	saldatura *f* a ultrasuoni
contre-dépouille *f*	contrasalida *f*; destalonado *m*	sottosquadra *f*

Nr.	English	Deutsch
1699	**underwater pelletiser**	Unterwassergranulator *m*
1700	**uniaxial**	einachsig
1701	**unplasticised**	weichmacherfrei
1702	**unsaturated**	ungesättigt
1703	**unsaturated polyester resin** *(UP)*	ungesättigtes Polyesterharz *n*
1704	**unscrewing core**	Ausschraubkern *m*
1705	**unscrewing device**	Ausdrehvorrichtung *f*
1706	**unscrewing mould**	Ausschraubwerkzeug *n*
1707	**unwind unit**	Abwickelvorrichtung *f*
1708	**urea**	Harnstoff *m*
1709	**urea resins**	Harnstoffharze *npl*
1710	**urethane**	Urethan *n*
1711	**UV absorber**	UV-Absorber *m*
1712	**UV curing**	UV-härtend
1713	**UV resistance**	UV-Beständigkeit *f*

Français	Español	Italiano
granulateur *m* sous l'eau	granuladora *f* bajo agua	estrusore *m* granulatore in acqua
uniaxial	uniaxial	monoassiale
non plastifié	libre de plastificantes	non plastificato
insaturé	insaturado	insaturo
résine *f* polyester insaturé	resina *f* poliéster insaturado	resina *f* poliestere insatura
noyau *m* à dévissage	núcleo *m* desenroscable	inserto *m* a svitamento
dispositif *m* de dévissage	mecanismo *m* para desenroscar	dispositivo *m* di svitamento
moule *m* à dévissage; moule *m* à tiroirs	molde *m* desenroscable	stampo *m* a svitamento
machine *f* à dérouler	dispositivo *m* de bobinado	unità *f* di svolgimento
urée *f*	urea *f*	urea *f*
résines *fpl* urées	resinas *fpl* de urea	resine *fpl* ureiche
uréthanne *m*	uretano *m*	uretano *m*
absorbeur *m* d'UV	absorbente *m* de UV	assorbitore *m* UV
réticulation aux UV	curado por UV	reticolazione *f* UV
résistance *f* aux UV	resistencia *f* a rayos UV	resistenza *f* agli UV

Nr.	English	Deutsch
1714	**UV stabiliser**	UV-Stabilisator *m*
1715	**vacuum calibration**	Vakuumkalibrierung *f*
1716	**vacuum casting**	Vakuumgießen *n*
1717	**vacuum forming**	Vakuumverformung *f;* Tiefziehen *n*
1718	**vacuum injection process**	Vakuum-Injektions-Verfahren *n*
1719	**vacuum pump**	Vakuumpumpe *f*
1720	**valve**	Ventil *n*
1721	**vapour deposition**	Aufdampfen *n*
1722	**vapour pressure**	Dampfdruck *m*
1723	**vent zone**	Entgasungszone *f*
1724	**vented extruder**	Entgasungsextruder *m*
1725	**vented hopper**	Entgasungstrichter *m*
1726	**vented screw**	Entgasungsschnecke *f*
1727	**venting**	Entgasen *n*

Français	*Español*	*Italiano*
stabilisant *m* UV	estabilizador *m* UV	stabillizzante *m* UV
calibrage *m* sous vide	calibrado *m* por vacío	calibrazione *f* sotto vuoto
coulée *f* sous vide	colada *f* por vacío	colaggio *m* sotto vuoto
formage *m* sous vide	moldeo *m* por vacío	formatura *f* sotto vuoto
procédé *m* d'injection sous vide	proceso *m* de inyección-vacío	processo *m* di iniezione sotto vuoto
pompe *f* à vide	bomba *f* de vacío	pompa *f* di depressione
clapet *m;* valve *f*	válvula *f*	valvola *f*
déposition *f* de vapeur	metalizado *m* por alto vacío	deposizione *f* di vapore
pression *f* de vapeur	presión *f* de vapor	pressione *f* di vapore
zone *f* de dégazage	zona *f* de desgasificación	zona *f* di ventilazione
extrudeuse *f* à dégazage	extrusora *f* con des gasificación	estrusore *m* con degasaggio
trémie *f* de dégazage	embudo *m* de vacío	tramoggia *f* con degasaggio
vis *f* de degazage	tornillo *m* con desgasificación	vite *f* con degasaggio
dégazage *m*	desgasificación *f*	degasaggio *m*

Nr.	English	Deutsch
1728	**venting channel**	Entlüftungskanal *m*
1729	**venting unit**	Entgasungseinheit *f*
1730	**vertical adjustment**	Höhenverstellung *f*
1731	**vertical flash face**	Tauchkante *f*
1732	**vibrating table**	Rütteltisch *m*
1733	**vibration stress**	Schwingungsbeanspruchung *f*
1734	**vibration welding**	Vibrationsschweißen *n*
1735	**Vicat softening temperature** *(VST)*	Vicat-Erweichungs-temperatur *f*
1736	**Vickers hardness**	Vickers-Härte *f*
1737	**vinyl chloride**	Vinylchlorid *n*
1738	**vinyl ester resin**	Vinylesterharz *n*
1739	**vinyl polymers**	Vinylpolymere *npl*
1740	**virgin material**	Neuware *f*
1741	**viscoelasticity**	Viskoelastizität *f*
1742	**viscometer**	Viskosimeter *m*

Français	Español	Italiano
trou *m* d'évent	salida *m* de aire; respiradero *m* de aire	sfogo *m* per l'aria
système *m* de dégazage	unidad *f* de desgasificación	unità *f* di degasaggio
réglage *m* vertical	ajuste *m* vertical	regolazione *f* verticale
plaque *f* d'échappement	canto *m* sumergido	faccia *f* a bava verticale
table *f* vibrante	mesa *f* vibratoria	tavola *f* vibrante
contrainte *f* de vibration	esfuerzo *m* de vibración	sollecitazione *f* da vibrazione
soudage *m* par vibration	soldeo *m* por vibración	saldatura *f* per vibrazione
point *m* Vicat	temperatura *f* de reblandecimiento Vicat	temperatura *f* di rammollimento Vicat
dureté *f* Vickers	dureza *f* Vickers	durezza *f* Vickers
chlorure *m* de vinyle	cloruro *m* de vinilo	cloruro *m* di vinile
résine *f* vinyl ester	resina *f* de vinil éster	resina *f* vinilestere
polymères *mpl* vinyliques	polímeros *mpl* vinílicos	polimeri *mpl* vinilici
matière *f* vierge	material *m* virgen	materiale *m* vergine
viscoélasticité *f*	viscoelasticidad *f*	viscoelasticità *f*
viscosimètre *m*	viscosímetro *m*	viscosimetro *m*

Nr.	English	Deutsch
1743	**viscosity**	Viskosität *f*
1744	**viscosity number**	Viskositätszahl *f*
1745	**viscous**	viskos
1746	**void**	Lunker *m*
1747	**volatile**	flüchtig
1748	**volatile matter**	Flüchte *f;* flüchtige Substanz *f*
1749	**volume resistance**	Durchgangswiderstand *m*
1750	**volumetric feeding**	Volumendosierung *f*
1751	**vulcanisation**	Vulkanisation *f;* Vernetzung *f*
1752	**wall-adhering**	wandhaftend
1753	**wall-slipping**	wandgleitend
1754	**wall thickness control**	Wanddickensteuerung *f*
1755	**warpage**	Verzug *m*
1756	**waste**	Abfall *m*

Français	*Español*	*Italiano*
viscosité *f*	viscosidad *f*	viscosità *f*
indice *m* de viscosité	número *m* de viscosidad	indice *m* di viscosità
visqueux	viscoso	viscoso
porosité *f*	poro *m;* hueco *m* interno	soffiatura *f;* risucchio *m*
volatil	volatil	volatile
substance *f* volatile	sustancia *f* volátil	sostanza *f* volatile
résistance *f* au courant de cheminement; résistance *f* volumique	resistividad *f* transversal	resistenza *f* di volume
dosage *m* volumétrique	dosificación *f* volumétrica	alimentazione *f* volumetrica
vulcanisation *f*	vulcanización *f;* reticulación *f*	vulcanizzazione *f*
adhérent à la paroi	pared adhesiva	adesione alla parete
glissant sur la paroi	pared resbaladiza	scivolositá *f* della parete
régulation *f* d'épaisseur de paroi	regulador *m* del espesor de pared	controllo *m* spessore parete
gauchissement *m*	deformación *f*	deformazione *f;* arriciamento *m*
déchet *m*	desperdicio *m*	rifiuti *mpl*

Nr.	English	Deutsch
1757	**water absorption**	Wasseraufnahme *f*
1758	**water** *(assisted)* **injection moulding**	Wasserinjektionsverfahren *n*
1759	**water content**	Wassergehalt *m*
1760	**water cooling**	Wasserkühlung *f*
1761	**water jet cutting**	Wasserstrahlschneiden *n*
1762	**water soluble**	wasserlöslich
1763	**water treeing**	Bäumchenbildung *f*
1764	**water vapour permeability**	Wasserdampfdurchlässigkeit *f*
1765	**wavelength**	Wellenlänge *f*
1766	**wax**	Wachs *n*
1767	**wear**	Verschleiß *m*
1768	**wear protection**	Verschleißschutz *m*
1769	**wear resistance**	Verschleißwiderstand *m*
1770	**wear resistant**	verschleißarm

Français	*Español*	*Italiano*
absorption *f* d'eau	absorción *f* de agua	assorbimento *m* d'acqua
moulage *m* par injection assisté par l'eau	moldeo *m* por inyección asistido por agua	stampaggio *m* a iniezione con acqua
contenu *m* en eau	contenido *m* de agua	contenuto *m* d'acqua
refroidissement *m* à l'eau	refrigeración *f* por agua	raffreddamento *m* ad acqua
découpage *m* au jet d'eau	corte *m* por chorro de agua	taglio *m* a getto d'acqua
soluble dans l'eau	soluble en agua	idrosolubile
arborescence *f*	arborescencia *f*	arborescenza *f*
perméabilité *f* à la vapeur d'eau	permeabilidad *f* al vapor de agua	permeabilità *f* al vapore d'acqua
longueur *f* d'onde	longitud *f* de onda	lunghezza *f* d'onda
cire *f*	cera *f*	cera *f*
usure *f*	abrasión *f*	usura *f*
protection *f* anti usure	protección *f* antidesgaste	protezione *f* antiusura
résistance *f* à l'usure	resistencia *f* a la abrasión	resistenza *f* all'usura
résistant à l'usure	resistente a la abrasión	resistente all'usura

Nr.	English	Deutsch
1771	**weather resistant**	witterungsbeständig
1772	**weathering**	Bewitterung *f*
1773	**weatherometer**	Bewitterungsgerät *n*
1774	**web**	Bahn *f*
1775	**web guide**	Bahnführung *f*
1776	**weight feeding**	Gewichtsdosierung *f*
1777	**weld line**	Schweißnaht *f;* Bindenaht *f*
1778	**weld strength**	Bindenahtfestigkeit *f;* Schweißnahtfestigkeit *f*
1779	**welding**	Schweißen *n*
1780	**welding machine**	Schweißanlage *f*
1781	**welding rod**	Schweißdraht *m*
1782	**wet, to**	benetzen
1783	**wettability**	Benetzbarkeit *f*
1784	**wetting**	Benetzen *n*

Français	*Español*	*Italiano*
résistant aux intempéries	resistente a la intemperie	resistente all'invecchiamento
exposer *f* aux intempéries	exposición *f* a la intemperie	invecchiamento *m*
appareil *m* de viellissement accéléré	equipo *m* de envejecimiento acelerado	apparecchiatura *f* per invecchiamento accelerato
bande *f* continue	lámina *f* continua; cinta *f* continua	nastro *m*; cinta *f* continua
guide *m* de bande	guía *f* de la lámina	guidanastro *m*
alimentation *f* en poids	alimentación *f* por peso	alimentazione *f* ponderale
ligne *f* de soudure	línea *f* de soldadura	linea *f* di saldatura
résistance *f* de la soudure	resistencia *f* de la linea de reunión; resistencia *f* de la linea de soldadura	resistenza *f* della saldatura
soudage *m*	soldadura *f*; soldeo *m*	saldatura *f*
soudeuse *f*	instalación *f* de soldeo	saldatrice *f*
cordon *m* de soudure	varilla *f* de soldar	barra *f* di saldatura
mouiller	mojar	bagnare
mouillabilité *f*	mojabilidad *f*	bagnabilità *f*
mouillant *m*	humectación *f*	umettazione *f*

Nr.	English	Deutsch
1785	**wetting agent**	Benetzungsmittel *n*
1786	**wind up, to**	wickeln; aufwickeln
1787	**wind-up unit; winder**	Wickelanlage *f*
1788	**winding process**	Wickelverfahren *n*
1789	**wire covering**	Drahtummantelung *f*
1790	**wood flour**	Holzmehl *n*
1791	**woven fabric**	Gewebe *n*
1792	**yellowing**	Gilbung *f*
1793	**yellowness index**	Vergilbungsindex *m*
1794	**yield point**	Fließgrenze *f;* Streckgrenze *f*
1795	**yield stress**	Streckspannung *f*

Français	*Español*	*Italiano*
agent *m* mouillant	agente *m* humectante	agente *m* umettante
enrouler; bobiner	enrollar; bobinar	avvolgere
enrouleur *m*	bobinadora *f*	avvolgitore *m*
enroulement *m*	proceso *m* de bobinado	processo *m* di avvolgimento
revêtement *m* de câble	recubrimiento *m* de hilos	rivestimento *m* cavi
farine *f* de bois	serrín *f* de madera	segatura *f*
tissu *m*	tejido *m*	tessuto *m*
jaunissement *m*	amarilleamiento *m*	ingiallimento *m*
indice *m* de jaunissement	índice *m* de amarilleamiento	indice *f* di ingiallimento
seuil *m* d'écoulement; limite *f* élastique	límite *m* de elasticidad	limite *m* di snervamento
tension *f* à la limite d'élasticité	esfuerzo *m* en punto de fluencia	sollecitazione *f* a snervamento

ENGLISH

DEUTSCH

FRANÇAIS

ESPAÑOL

ITALIANO

Technical Information for the CD-ROM

System Requirements:
Operating systems: Windows 95/98, WinNT 4.0, Windows 2000.
Amount of RAM: Recommended 32 Mbyte. CD-ROM Drive:
Recommended 4xspeed or higher.

Installations Instructions:
Explorer: Choose your CD-ROM Directory and double-click
"Setup.exe" from the root of your CD-ROM drive. Follow the
instructions of the Setup-Wizard.

Starting Glossary:
Setup creates a program group. If you didn't change the default
value "Carl Hanser Verlag" during installation this is where you
will find a programm link called "Glossary". Double-Click on this
link to start Glossary. Please note that Glossary runs only with this
CD inserted in your local CD-ROM drive.

This mini-CD runs on all commercially available CD-ROM drives.